AF552502

UNDERSTANDING GENETICS AND EVOLUTION

UNDERSTANDING GENETICS AND EVOLUTION

By

Dr. Rajiv Tyagi

Department of Zoology

M.M. College

Modi Nagar (U.P.)

(India)

DISCOVERY PUBLISHING HOUSE PVT. LTD.

NEW DELHI-110 002

First Published-2009

ISBN 978-81-8356-459-5

Published by:

DISCOVERY PUBLISHING HOUSE PVT. LTD.

4831/24, Ansari Road, Prahlad Street,
Darya Ganj, New Delhi-110002 (India)
Phone: 23279245 • Fax: 91-11-23253475
E-mail: dphbooks@rediffmail.com
dphtemp@indiatimes.com
Website: www.discoverypublishinghouse.com

Printed at:

Sachin Printers
Delhi

Preface

The present title "Understanding Genetics and Evolution" has been written for those students interested in careers in diverse fields of biological sciences. It provides a structured approach to learning by covering all the important topics in a uniform, systematic format. The book has been comprehensively designed incorporating recent advances in this fast moving field. It also provides accessible information on molecular biology in compact form for undergraduate students in biology and related life sciences. It is intelligible to the educated layman, though it deals with some complex ideas. It is an adequate text for all the requirements of students in this area. In addition, busy lecturers who require a quick reference compendium will find it useful, particularly for tutional planning. Simple, yet hopefully clear figures and tables are provided throughout the book.

The over-riding goal of this book, and indeed of the whole *Understanding series,* is to present the essential information concering molecular biology in a compact, readily accessible form which leads itself to student learning and revision. The convergence of various approaches has generated a rich panorama of detail, the significance of which we are still attempting to unraval. The present text has been written as an introduction to this rapidly growing field.

To make the work more comprehensive and informative, the author has consulted many authoritative books, research journals, abstracts, monographs etc., so there can be no claim to originality except in the manner of treatment.

The author expresses his thanks to his friends and colleagues whose continue inspirations have initiated him to bring out this book.

The author expresses his gratitude to Mr. Wasan and staff of M/s Discovery Publishing House Pvt. Ltd. for their whole hearted cooperation in the publication of this book.

In the mean time, the author will remain sincerely responsible for any shortcomings of the book and be grateful to the readers for their suggestions and constructive criticism for the continuous betterment of the book. He takes this opportunity to appeal to the readers to send their suggestions straightaway to his Publisher.

Author

Contents

1

Origin of Molecules

How many of the 90 naturally occurring elements are essential to life? After more than a century of increasingly refined investigation, the question still cannot be answered with certainty. Only a year or so ago the best answer would have been 20. Since then four more elements have been shown to be essential for the growth of young animals: fluorine, silicon, tin and vanadium. Nickel may soon be added to the list. In many cases the exact role played by these and other trace elements remains unknown or unclear. These gaps in knowledge could be critical during a period when the biosphere is being increasingly contaminated by synthetic chemicals and subjected to a potentially harmful redistribution of salts and metal ions. In addition, new and exotic chemical form of metals (such as methyl mercury) are being discovered, and a complex series of competitive and synergistic relations among mineral salts has been encountered. We are led to the realization that we are ignorant of many basic facts about how our chemical milieu affects our biological fate. Biologists and chemists have long been fascinated by the way evolution has selected certain elements as the building blocks of living organisms and has ignored others.

The composition of the earth and its atmosphere obviously sets a limit on what elements are available. The earth itself is hardly a chip off the universe. The solar system, like the universe, seems to be 99 per cent hydrogen and helium. In the earth's crust helium is essentially nonexistent (except in a few rare deposits) and hydrogen atoms constitute only about .22 per cent of the total. Eight elements provide more than 98 per cent of the atoms in the earth's crust: oxygen (47 per cent), silicon (28 per cent), aluminium (7.9 per cent), iron (4.5 per cent), calcium (3.5 per cent), sodium (2.5 per cent), potassium (2.5 per cent) and magnesium (2.2 per cent). Of these eight elements only five are among the 11 that account for more than 99.9 per cent of the atoms in the human body. Not surprisingly nine of the 11 are also the nine most abundant elements in sea-water. Two elements, hydrogen and oxygen, account for 88.5 per cent of the atoms in the human body; hydrogen supplies 63 per cent of the total and oxygen 25.5 per cent. Carbon accounts for another 9.5 per cent and nitrogen 1.4 per cent. The remaining 20 elements now thought to be essential for mammalian life account for less than .7 per cent of the body's atoms.

Background of Selection

Three characteristics of the biosphere or of the elements themselves appear to have played a major part in directing the chemistry of living forms. *First and foremost* there is the ubiquity of water, the solvent base of all life on the earth. Water is a unique compound; its stability and boiling point are both unusually high for a molecule of its simple composition. Many of the other compounds essential for life derive their usefulness from their response to water fullness from their response to water: whether they are soluble or insoluble, whether or not (if they are soluble) they carry an electric charge in solution and, not least, what effect they have on the viscosity of water. The *second directing force* involves the chemical properties of carbon, which evolution selected over silicon as the central building block for constructing giant molecules. Silicon is 146 times more plentiful than carbon in the earth's crust and exhibits many of the same properties. Silicon is directly below carbon in the periodic table of the elements; like carbon, it has the capacity to gain four electrons and form four covalent bonds.

The crucial difference that led to the preference for carbon compounds over silicon compounds seems traceable to two chemical features: the unusual stability of carbon dioxide, which is readily soluble in water and always monometric (it remains a single molecule), and the almost unique ability of carbon to form long chains and stable rings with five or six members. This versatility of the carbon atom is responsible for the millions of organic compounds found on the earth. *Silicon* in contrast, is insoluble in water and forms only relatively short chains with itself. It can enter into longer chains, however, by forming alternating bonds with oxygen, creating the compounds known as silicones (-Si-O-Si-O-Si). Carbon-to-carbon bonds are more stable than silicon-to-silicon bonds, but not so stable as to be virtually immutable, as the silicon-oxygen polymers are. Nevertheless, silicon has recently been shown to be essential in a way as yet unknown for normal bone development and full growth in chicks.

The third force influencing the evolutionary selection of the elements essential for life is related to an atom's size and charge density. Obviously the heavy synthetic elements from neptunium (atomic number 93) to lawrencium (No. 103), along with two lighter synthetic elements, technetium (No. 43) and promethium (No. 61), were never available in nature. (The atomic number expresses the number of protons in the nucleus of an atom or the number of electrons around the nucleus). The eight heavy elements in another group (No. 84 and 85 and Nos. 87 through 92) are too radioactive to be useful in living structures. Six more elements are inert gases with virtually no useful chemical reactivities: helium, neon, argon, krypton, xenon and radon.

On various plausible grounds one can exclude another 24 elements, or a total of 38 natural elements, as being clearly unsatisfactory for incorporation in living organisms because of their relative unavailability (particularly the elements in the lanthanide and actinide series) or their high toxicity (for example mercury and lead). This leaves 52 of the 90 natural elements as being potentially useful. Only three of the 24 elements known to be essential for animal life have an atomic number above 34. All three are needed only in trace amounts: molybdenum (No. 42), in (No. 50) and iodine (No. 53). The four most abundant atoms in living organisms—hydrogen, carbon, oxygen and nitrogen—have atomic numbers of 1, 6, 7 and 8. Their preponderance seems attributable to their being the smallest and lightest elements that can achieve stable electronic configurations by adding one to four electrons.

The ability to add electrons by sharing them with other atoms is the first step in forming chemical bonds leading to stable molecules. The seven next most abundant elements in living organisms all have atomic numbers below 21. In the order of their abundance in mammals they are calcium (No. 20), phosphorus (No. 15), potassium (No. 19), sulphur (No. 16), sodium (No. 11), magnesium (No. 12) and chlorine (No. 17). The remaining 10 elements known to be present in either plants or animals are needed only in traces. With the exception of fluorine (No. 9) and silicon (No. 14), the remaining eight occupy positions between No. 23 and No. 34 in the periodic table.

Table 1.1. Comparison between the composition of the human body with the approximate composition of seawater, the earth's crust and the universe at large.

Composition of Universe		*Composition of Earth's Crust*		*Composition of Seawater*		*Composition of Human Body*	
Percent of Total Number of Atoms							
H	91	O	47	H	66	H	63
He	9.1	Si	28	O	33	O	25.5
O	.057	Al	7.9	Cl	.33	C	9.5
N	.042	Fe	4.5	Na	.28	N	1.4
C	.021	Ca	3.5	Mg	.033	Ca	.31
Si	.003	Na	2.5	S	.017	P	.22
Ne	.003	K	2.5	Ca	.006	Cl	.03
Mg	.002	Mg	2.2	k	.006	K	.06
Fe	.002	Ti	.46	C	.0014	S	.05
S	.001	H	.22	Br	.0005	Na	.03
		C	.09			Mg	.01
All others <.01		All others <.1		All others <.1		All others <.01	

It is interesting that this interval embraces three elements for which evolution has evidently found no role: gallium, germanium and arsenic. None of the metals with properties similar to those of gallium (such as aluminum and indium) has proved to be useful to living organisms. On the other hand, since silicon and tin, two elements with chemical activities similar to those of germanium, have just joined the list of essential elements, it seems possible that germanium too, in spite of its rarity, will turn out to have an essential role. Arsenic, of course, is a well-known poison.

Functions of Essential Elements

Some useful generalizations can be made about the role of the various elements. Six elements—carbon, nitrogen, hydrogen, oxygen, phosphorus and sulphur—make up the molecular building blocks of living matter: amino acids, sugars, fatty acids, purines, pyrimidines and nucleotides. These molecules not only have independent biochemical roles but also are the respective constituents of the following large molecules: proteins, glycogen, starch, lipids and nucleic acids. Several of the 20 amino acids contain sulphur in addition to carbon, hydrogen and oxygen.

Phosphorous plays an important role in the nucleotides such as adenosine triphosphate (ATP), which is central to the energetics of the

Table 1.2. 21 out of the first 34 elements in the periodic table which are essential for animal life.

Element	*Symbol*	*Atomic Number*	*Comments*
Hydrogen	H	1	Required for water and organic compounds.
Helium	He	2	Inert and unused.
Lithium	Li	3	Probably unused.
Beryllium	Be	4	Probably unused; toxic.
Boron	B	5	Essential in some plants, function unknown.
Carbon	C	6	Required for organic compounds.
Nitrogen	N	7	Required for many organic compounds.
Oxygen	O	8	Required for water and organic compounds.
Fluorine	F	9	Growth factor in rats; possible constituent of teeth and bone.
Neon	Ne	10	Inert and unused
Sodium	Na	11	Principal extracellular cation.
Magnesium	Mg	12	Required for activity of many enzymes in chlorophyll.
Aluminium	Al	13	Essentiality under study.
Silicon	Si	14	Possible structural unit of diatoms; recently shown to be essential in chicks.
Phosphorous	P	15	Essential for biochemical synthesis and energy transfer.
Sulphur	S	16	Required for proteins and other biological compounds.
Chlorine	Cl	17	Principal cellular and extracellular anion.
Argon	A	18	Inert and unused.
Potassium	K	19	Principal cellular cation.
Calcium	Ca	20	Major component of bone; required for some enzymes.
Scandium	Sc	21	Probably unused.
Titanium	Ti	22	Probably unused.
Vanadium	V	23	Essential in lower plants, certain marine animals and rat.
Chromium	Cr	24	Essential in higher animals; related to action of insulin.
Manganese	Mn	25	Required for activity of several enzymes.

Iron	Fe	26	Most important transition metal ion essential for hemoglobin and many enzymes.
Cobalt	Co	27	Required for activity of several enzymes in vitamin B_{12}.
Nickel	Ni	28	"Essentiality under study.
Copper	Cu	29	Essential in oxidative and other enzymes and hemocyanin.
Zinc	Zn	30	Required for activity of many enzymes.
Gallium	Ca	31	Probably unused.
Germanium	Ge	32	Probably unused.
Arsenic	As	33	Probably unused; toxic
Selenium	Se	34	Essential for liver function
Molybdenum	Mo	42	Required for activity of several enzymes.
Tin	Sn	50	Essential in rats; function unknown.
Iodine	I	53	Essential constituent of the thyroid hormones.

cell. ATP includes components that are also one of the four nucleotides needed to form the double helix of deoxy-ribonucleic acid (DNA), which incorporates the genetic blueprint of all plants and animals. Both sulphur and phosphorous are present in many of the small accessory molecules called coenzymes. In bony animals phosphorous and calcium help to create strong supporting structures. The electrochemical properties of living matter depend critically on elements or combinations of elements that either gain or lose electrons when they are dissolved in water, thus forming ions.

The principal cations (electron-deficient, or positively charged, ions are provided by four metals: sodium, potassium, calcium and magnesium. The principal anions (ions with a negative charge because they have surplus electrons) are provided by the chloride ion and by sulphur and phosphorous in the form of sulfate ions and phosphate ions. These seven ions maintain the electrical neutrality of body fluids and cells and also play a part in maintaining the proper liquid volume of the blood and other fluid systems. Whereas the cell membrane serves as a physical barrier to the exchange of large molecules, it allows small molecules to pass freely.

The electrochemical functions of the anions and cations serve to maintain the appropriate relation of osmotic pressure and charge distribution on the two sides of the cell membrane. One of the striking features of the ion distribution is the specificity of these different ions. Cells are rich in potassium and magnesium, and the surrounding plasma is rich in sodium and calcium. It seems likely that the distribution of ions in the plasma of higher animals reflects the oceanic origin of their evolutionary antecedents. One would like to know how primitive cells learned to exclude the sodium and calcium ions in which they were bathed and to develop an internal milieu enriched in potassium and magnesium. The third and last group of essential elements consists of the trace elements. The fact that they are required in extremely minute quantities in no way diminishes their great importance.

In this sense they are comparable to the vitamins. We now know that the great majority of the trace elements, represented by metallic ions, serve chiefly as key components of essential enzyme systems or of proteins with vital functions (such as hemoglobin and myoglobin, which respectively transports oxygen in the blood and stores oxygen in muscle). The heaviest essential element, iodine, is an essential constituent of the thyroid hormones thyroxine and triiodothyronine, although its precise role in hormonal activity is still not understood.

Trace Elements

To demonstrate that a particular element is essential to life becomes increasingly difficult as one lowers the threshold of the amount of a substance recognizable as a "trace." It has been known for more than 100 years, for example, that iron and iodine are essential to man. In a rapidly developing period of biochemistry between, 1928 and 1935 four more elements, all metals, were shown to be essential: copper, manganese, zinc and cobalt. The demonstration can be credited chiefly to a group of investigators at the University of Wisconsin led by C.A. Elvehjem, E.B. Hart and W.R. Todd. At that time it seemed that these four metals might be the last of the essential trace elements.

In the next 30 years, however, three more elements were shown to be essential chromium, selenium and molybdenum. Fluorine, silicon, tin and vanadium have been added since 1970. The essentiality of five of these last seven elements was discovered through the careful, paintstaking efforts of *Klaus Schwarz* and his associates, initially located at the National Institute of Health and now based at the Veterans Administration Hospital in Long Beach, Calif. For the past 15 years Schwarz's group has made a systematic study of the trace-element requirements of rates and other small animals. The animals are maintained from birth in a completely isolated sterile environment.

The apparatus is constructed entirely of plastics to eliminate the stray contaminants contained in metal, glass and rubber. Although even plastics may contain some trace elements, they are so tightly bound in the structural lattice of the material that they cannot be leached out or be picked up by an animal even through contact. A typical isolator system houses 32 animals in individual acrylic cages. Highly efficient air filters remove all trace substances that might be present in the dust in the air. Thus the animals only access to essential nutrients is through their diet. They receive chemically pure amino acids instead of natural proteins, and all other dietary ingredients are screened for metal contaminants.

Since the standards of purity employed in these experiments far exceed those for reagents normally regarded as analytically pure, Schwarz and his co-workers have had to develop many new alalytical chemical methods. The most difficult problem turned out to be the purification of salt mixtures. Even the purest commercial reagents were contaminated with traces of metal ions. It was also found that trace elements could be passed from mothers to their offspring. To minimize this source of contamination animals are weaned as quickly as possible, usually from 18 to 20 days after birth. With these precautions Schwarz and his colleagues have within the past several years been able to produce a new deficiency disease in rates. The animals grow poorly, lose hair and muscle tone, develop shaggy fur and exhibit other detrimental changes. When standard laboratory food is given these animals, they regain their normal appearance. At first it was thought that all the symptoms were caused by the lack of one particular trace element. Eventually four different elements had to be supplied to complete the highly purified diets the animals had been receiving. The four elements proved to be fluorine, silicon, tin and vanadium. A convenient source of these elements is yeast ash or liver preparations from a healthy animal. The animals on the deficiency diet grew less than half as fast as those on a normal or supplemented diet. Growth alone, however, may not tell the entire story. There is some evidence that even the addition of the four elements may not reverse the loss of hair and skin changes resulting from the deficiency diet.

Functions of Trace Elements

The addition of tin and vanadium to the list of essential trace metals brings to 10 the total number of trace metals needed by animals and plants. What role do these metals play? For six of the eight trade metals recognized from earlier studies (that is, for iron, zinc, copper, cobalt, manganese and molybdenum) we are reasonably sure of the answer. The six are constituents of a wide range of enzymes that participate in a variety of metabolic processes. In addition to its role in hemoglobin and myoglobin, iron appears in succinate dehydrogenase, one of the enzymes needed for the utilization of energy from sugars and starches.

Enzymes incorporating zinc help to control the formation of carbon dioxide and the digestion of proteins. Copper is present in more than a dozen enzymes, whose roles range from the utilization of iron to the pigmentation of the skin. Cobalt appears in enzymes involved in the synthesis of DNA and the metabolism of amino acids. Enzymes incorporating managnese are involved in the formation of urea and the metabolism of pyruvate. Enzymes incorporating molybdenum participate in purine metabolism and the utilization of nitrogen. These six meals belong to a group known as transition elements. They owe their uniqueness to their ability to form strong complexes with ligands, or molecular groups, of the type present in the side chains of proteins. Enzymes in which transition metals are tightly incorporated are called metalloenzymes, since the metal is usually embedded deep inside the structure of the protein. If the metal atom is removed, the protein usually loses its capacity to function as an enzyme.

There is also a group of enzymes in which the metal ion is more loosely associated with the protein but is nonetheless essential for the enzyme's activity. Enzymes in this group are known as metal-ion-activiated enzymes. In either group the role of the metal ion may be to maintain the proper conformation of the protein, to bind the substrate (the molecule acted on) to the protein or to donate or accept electrons in reactions where the substate is reduced or oxydized. In 1968 the complete three dimensional structure of the first metalloenzyme, cytochrome *c*, was published. Cytochrome *c,* a red enzyme containing iron, is universally present in plants and animals. It is one of a series of enzymes, all called cytochromes, that extract energy from food molecules by the stepwise addition of oxygen. The complete amino acid sequence of cytochrome *c* obtained from the human heart was determined some 10 years ago by a group led by *Emil L. Smith* of the University of Califormia at Los Angeles and by *Emanuel Margoliash* of North-western University.

The iron atom is partially complexed with an intricate organic molecule, protoporphyrin, to form a heme group similar to that in

Table 1.3. Wide variety of metalloenzymes required for the successful functioning of living organisms.

Metal	*Enzyme*	*Biological Function*
Iron	Ferredoxin	Photosynthesis
	Succinate Dehydrogenase	Aerobic oxidation of carbohydrates
Iron in Heme	Aldehyde Oxidase	Aldehyde oxidation
	Cytochromes	Electron transfer
	Catalase	Protection against hydrogen peroxide
	(Hemoglobin)	Oxygen transport
Copper	Ceruloplasmin	Iron utilization
	Cytochrome Oxidase	Principal terminal oxidase
	Lysine Oxidase	Elasticity of aortic walls
	Tyrosinase	Skin pigmentation
	Plastocyanin	Photosynthesis
	(Hemocyanin)	Oxygen transport in invertebrates
Zinc	Carbonic Anhydrase	CO_2 formation; regulation of acidity
	Carboxypeptidase	Protein digestion
	Alcohol Dehydrogenase	Alcohol metabolism
Manganese	Arginase	Urea formation
	Pyruvate Carboxylase	Pyruvate metabolism
Cobalt	Ribonucleotide Reductase	DNA biosynthesis
	Glutamate Mutase	Amino acid metabolism
Molybdenum	Xanthine Oxidase	Purine metabolism
	Nitrate Reductase	Nitrate utilization
Calcium	Lipase	Lipid digestion
Magnesium	Hexokinase	Phosphate transfer

hemoglobin. Of the iron atom's six coordination sites, four are attached to the heme group through nitrogen atoms. The other two sites form bonds with the protein chain; one bond is through a nitrogen atom in the side chain of a histidine unit at site No. 18 in the protein sequence and the other bond is through a sulphur atom in the side chain of a methionine unit at site No. 80. Although the cytochrome *c* molecule is complicated, it is one of the simplest of the metalloenzymes.

Cytochrome oxidase, probably the single most important enzyme in most cells, since it is responsible for transferring electrons to oxygen to form water, is far more complicated. Each molecule contains about 12 times as many atoms as cytochrome *c,* including two copper atoms and two heme groups, both of which participate in transferring the elections. More complicated yet is cysteamine oxygenase, which catalyzes the addition of oxygen to a molecule of cysteamine; it contains one atom each of three different metals: iron, copper and zinc.

There are many other combinations of metal ions and unique molecular assemblies. An extreme example is xanthine oxidase, which contains eight iron atoms, two molybdenum atoms and two molecules incorporating riboflavin (one of the B vitamins) in a giant molecule more than 25 times the size of cytochrome *c.* The metal-containing proteins of another group, the metalloproteins, closely resemble the metalloenzymes except that they lack an obvious catalytic function. Hemoglobin itself is an example. Others are hemocyanin, the copper-containing blue protein that carries oxygen in many invertebrates, metallothionein, a protein involved in the absorption and storage of zinc, and transferrin, a protein that transports iron in the bloodstream. There may be many more such compounds still unrecognized because their function has escaped detection.

Newest Essential Elements

Much remains to be learned about the specific biochemical role of the most recently discovered essential elements. In 1957 Schwarz and Calvin M. Foltz. working at the National Institutes of Health, showed that selenium helped to prevent several serious deficiency diseases in different animals, including liver necrosis and muscular dystrophy. Rats were protected against death from liver necrosis by a diet containing one-tenth of a part per million of selenium. Comparably low doses reversed the white muscle disease observed in cattle and sheep that happen to graze in areas where selenium is scarce.

In April a group at the University of Wisconsin under *J.T. Rotruck* reported a direct biochemical role for selenium. Oxidative damage to red blood cells was detected in rats kept on a selenium-deficient diet. This damage was related to reduced activity of an enzyme, glutathione peroxidase, that helps to protect hemoglobin against the injurious oxidative effects of hydrogen peroxide. The enzyme uses hydrogen peroxide to catalyze the oxidation of glutathione, thus keeping hydrogen peroxide from oxidizing the reduced state of iron in hemoglobin. Oxidized glutathione can readily be converted to reduced glutathione by a variety of intracellular mechanisms. There is some reason to believe glutathione peroxidase may even contain some form of selenium acting as an integral part of the functional enzyme molecule.

The physiological importance of chromium was established in 1959 by *Schwarz* and *Walter Mertz*. They found that chromium deficiency is characterized by impaired growth and reduced life-span, corneal lesions and a defect in sugar metabolism. When the diet is deficient in chromium, glucose is removed from the bloodstream only half as fast as it is normally. In rats the deficiency is relieved by a single administration of 20 micrograms of certain trivalent chromic salts. It now appears that the chromium ion works in conjunction with insulin, and that in at least some cases diabetes may reflect faulty chromium metabolism.

After developing the all-plastic trace-element isolator described above, *Schwarz, David B. Milne* and *Elizabeth Vineyard* discovered that tin, not previously suspected as being essential, was necessary for normal growth. Without one or two parts per million of tin in their diet, rats grow at only about two-thirds the normal rate. The next element shown to be essential in mammals by the Schwarz group was vanadium, an element that had been detected earlier in certain marine invertebrates but whose essentiality had not been demonstrated. On a diet in which vanadium is totally excluded rats suffer a retardation of about 30 per cent in growth rate. *Schwarz* and *Milne* found that normal growth is restored by adding one-tenth of a part per million of vanadium to the diet.

At higher concentrations vanadium is known to have several biological effects, but its essential role in trace amounts remains to be established. A high dose of vanadium blocks the synthesis of cholesterol and reduces the amount of phospholipid and cholesterol in the blood. Vanadium also promotes the mineralization of teeth and is effective as a catalyst in the oxidation of many biological substances. The third element most recently identified as being essential is fluorine. Even with tin and vanadium added to highly purified diets containing all other elements known to be essential, the animals in Schwarz's plastic cages still failed to grow at a normal rate. When up to half a part per million of potassium fluoride was added to the diet, the animals showed a 20 to 30 per cent weight gain in four weeks.

Although it had appeared that a trace amount of fluorine was essential for building sound teeth, Schwarz's study showed that fluorine's biochemical role was more fundamental than that. In any case fluoridated water provides more than enough fluorine to maintain a normal growth rate. Although there were earlier clues that silicon might be an essential life element, firm proof of its essentiality, at least in young chicks, was reported only three months ago. *Edith M. Carlisle* of the School of Public Health at the University of California at Los Angeles finds that chicks kept on a silicon-free diet for only one or two weeks exhibit poor development of feathers and skeleton, including markedly thin leg bones. The addition of 30 parts per million of silicon to the diet increases the chicks' growth more than 35 per cent and makes possible normal feathering and skeletal development.

Considering that silicon is not only the second most abundant element in the earth's crust but is also similar to carbon in many of its chemical properties, it is hard to see how evolution could have totally excluded it from an essential biochemical role. Nickel, nearly always associated with iron in natural substances, is another element receiving close attention. Also a transition element, it is particularly difficult to remove from the food used in special diets. Nickel seems to influence the growth of wing and tail feathers in chicks but more consistent data are needed to establish its essentiality. One incidental result of Schwarz's work has been the discovery of a previously unrecognized organic compound, which will undoubtedly prove to be a new vitamin.

Synergism and Antagonism

The interaction of the various essential metals can be extremely complicated. The absence of one metal in the diet can profoundly influence, either positively or negatively, the utilization of another metal that may be present. For example, it has been known for nearly 50 years that copper is essential for the proper metabolism of iron. An animal deprived of copper but not iron develops anaemia because the biosynthetic machinery fails to incorporate iron in hemoglobin molecules. It has only recently been found in our laboratories at Florida State University that ceruloplasmin, the copper-containing protein of the blood, is a direct molecular link between the two metals. Ceruloplasmin promotes the release of iron from animal liver so that the iron-binding protein of the serum, transferring, can complex with iron and transfer it to the developing red blood cells for direct utilization in the biosynthesis of hemoglobin. This represents a synergistic relation between copper and iron.

As an example of antagonism between elements one can cite the instance of copper and zinc. The ability of sheep or cattle to absorb copper is greatly reduced if too much zinc or molybdenum is present in their diet. Evidently either of the two metals can displace copper in an absorption process that probably involves competition for sites on a metal-binding protein in the intestines and liver.

The recent discoveries present many fresh challenges to biochemists. One can expect the discovery of previously unsuspected metalloenzymes containing vanadium, tin, chromium and selenium. New compounds or enzyme systems requiring fluorine and silicon may also be uncovered. The multiple and complex interdependencies of the elements suggest many hitherto unrecognized and important facts about the role and inter-relations of metal ions in nutrition and in health and disease.

2

ORIGIN OF LIFE

The problem of how life could have originated by ordinary chemical means long seemed insuperable. This is hardly surprising if we consider the present complexity of even the most elementary unit of life, the cell. At its simplest, a modern cell is surrounded by a highly selective permeable membrane, composed of lipids and proteins, that regulates the kinds of substances that pass through. Within the cell, the cytoplasm consists of a multitude of structures and substructures involved in the synthesis, storage, and breakdown of a large variety of chemical compounds.

AMINO ACIDS

Foremost among the metabolic agents that enable the cell to function are the many different proteins that catalyze and regulate practically all living chemical reactions. In basic structure, proteins consist of subunits, called *amino acids*, that have the following features:

```
      H  O
      |  |
H2—C*—C—OH
      |
      R
```

- An alpha carbon atom (C*) to which all other parts are attached
- An amino NH_2 group with a potential positive charge (NH_3^+)
- A carboxyl COOH group with a potential negative charge (COO^-)
- An H atom
- An R side chain that varies in structure among the different amino acids

These amino acids link together by chemical bonds called *peptide linkages* into linear *polypeptide chains* that are the constituents of proteins. The highly specific structure of any protein molecule, whether it functions as an *enzyme* (catalyst) or for some other purpose, derives from the exact linear placement of its various component amino acids. These specific amino acid sequences enable polypeptide chains to fold into specific three-dimensional forms that confer specific properties on proteins. It is, in fact, reasonable to claim that most present living phenomena, whether absorption, sensation, motion, structure, or whatever, derive from the enzymatic and regulatory activities of these long sequences of amino acids. Complexity, however, is not limited to proteins, since the amino acid sequences of proteins are actually determined by the nucleotide sequences in another group of basic molecules, nucleic acids.

Nucleic Acids

Nucleic acids are long-chained molecules composed of *nucleotide* subunits, each containing a pentose (5-carbon) sugar, a monophosphate group, and a nitrogenous base. The two kinds of sugar used in nucleic acids, ribose (hydroxylated at the

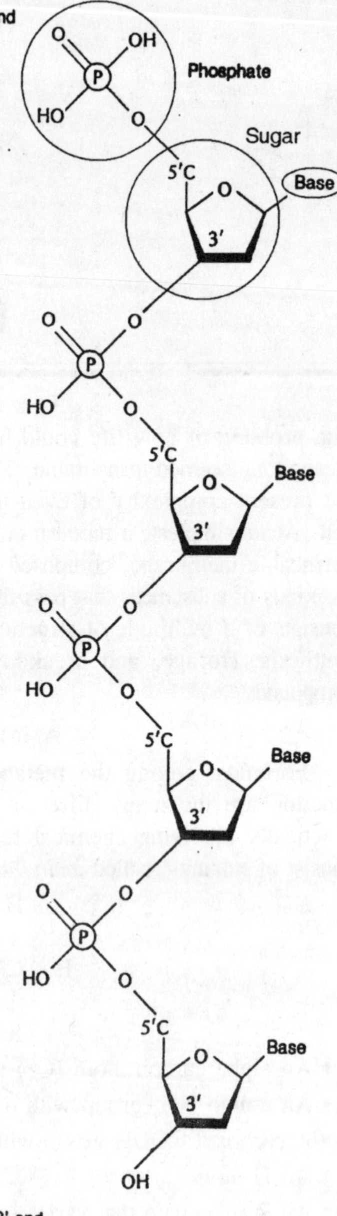

Fig. 2.1. General structure for DNA and RNA chain

2' carbon position) and deoxyribose (lacks 2' hydroxyl) provide the names for the two kinds of nucleic acids, ribonucleic acid (RNA) and deoxyribonucleic acid (DNA). In both these nucleic acids, the phosphate groups occupy the same position, tying the 3' carbon of one sugar to the 5' carbon of its neighbour via a phosphodiester bond. Connected to the 1' carbon of each sugar is one of four kinds of nitrogenous heterocyclic bases, of which two are purines (A, adenine, and G, guanine, in both DNA and RNA) and two are pyrimidines (C, cytosine, and T, thymine, in DNA and C, cytosine, and U, uracil, in RNA).

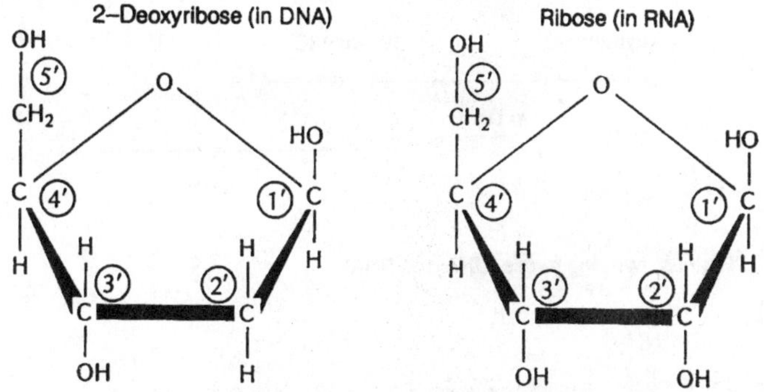

Fig. 2.2. Difference between the sugars found in DNA (deoxyribose) and RNA (ribose).

Since the complexity of proteins derives from the complexity of nucleic acids, you might think that the restriction of nucleic acid composition to only four different kinds of bases would limit the message-bearing capacity of these molecules to only four kinds of messages, but it doesn't. The facts that nucleic acid molecules may be many thousands or millions of nucleotides long, and that each message can be encoded by a unique linear sequence of nucleotides, endow these molecules with the capacity to bear an immense variety of highly complex messages. For any one nucleotide position, 4 different messages are possible (A, G, C, or T); for two nucleotides in tandem, 4^2 or 16 different messages are possible (AA, AG, AC, AT, GG, GC,...); and so on: the rule being simply that for a linear sequence of n nucleotides, 4^n different possible messages can be encoded. Thus a linear sequence of only 10 nucleotides can discriminate among more than 1 million (4^{10}) potentially different messages.

All this helps explain the information-carrying role of nucleic acids but does not explain how they replicate and transmit this information. The model presently accepted for nucleic acid replication derives from

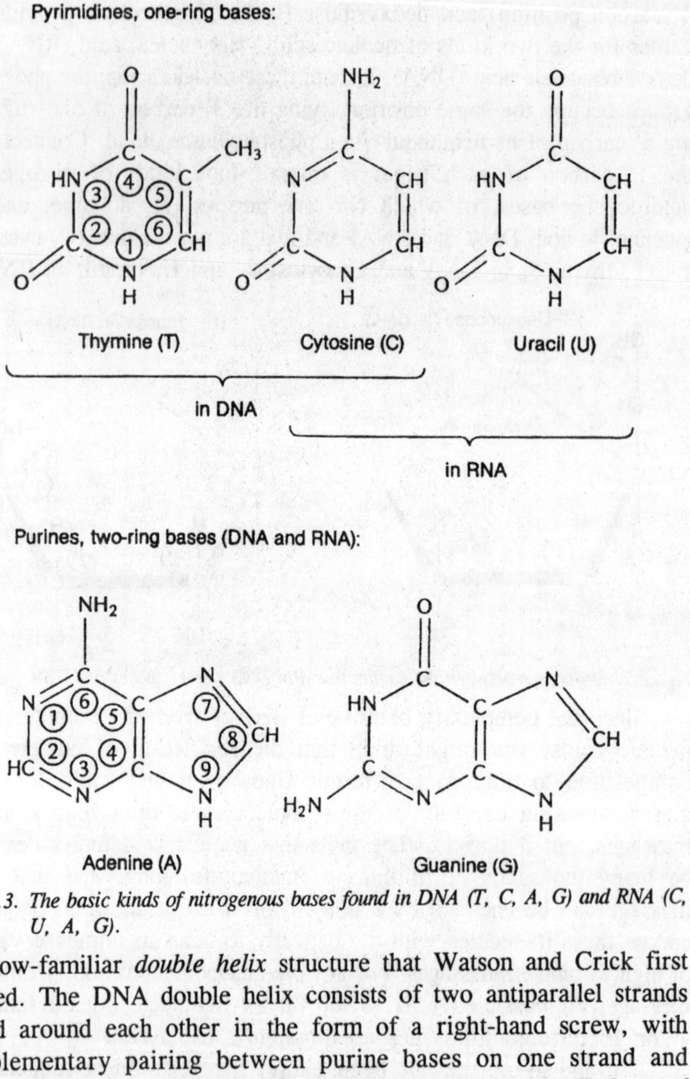

Pyrimidines, one-ring bases:

Thymine (T) Cytosine (C) Uracil (U)

in DNA

in RNA

Purines, two-ring bases (DNA and RNA):

Adenine (A) Guanine (G)

Fig. 2.3. The basic kinds of nitrogenous bases found in DNA (T, C, A, G) and RNA (C, U, A, G).

the now-familiar *double helix* structure that Watson and Crick first offered. The DNA double helix consists of two antiparallel strands coiled around each other in the form of a right-hand screw, with complementary pairing between purine bases on one strand and pyrimidine bases on the other (A-T, G-C). In the familiar B form of DNA, there are approximately 10 base pairs for each complete turn of the helix, and the bases are stacked almost perpendicularly to the helical axis.

The replicatory power of the DNA double helix obviously derives from the ability of each of the two strands to serve as a template for a newly complementary strand, so that two new double helices can

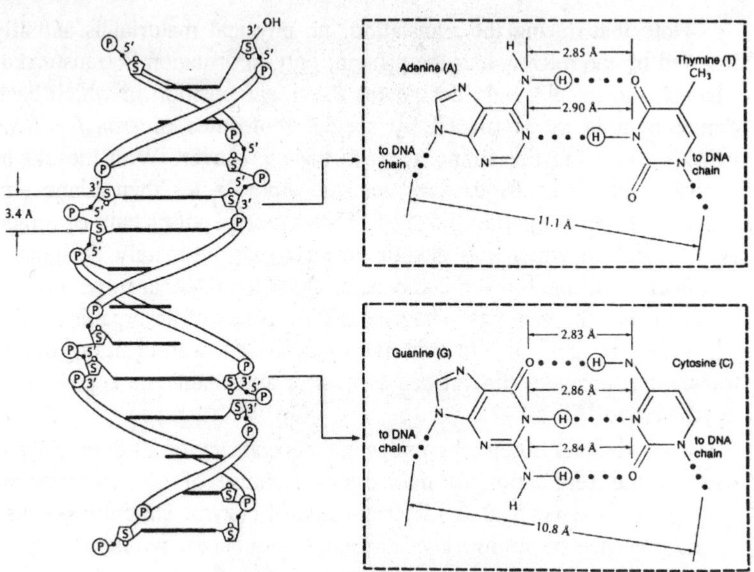

Fig. 2.4. The Watson-Crick model of the standard DNA double helix on the left, with examples of hydrogen-bond pairing between bases on the right (A–T, G–C).

form bearing nucleotide sequences that are identical to each other as well as to that of the parental molecule. This unique quality of exact molecular replication, enabling similar messages to be transmitted from generation to generation, confers on nucleic acids their function as "genetic material."

Fundamental to our understanding of the relationship between genetic material and protein, therefore, is an important concept: the three-dimensional structure of a protein—its form, shape, and subsequent function—is primarily determined by the linear sequence of amino acids of which it consists. This linear sequence of amino acids in turn derives from the linear sequence of bases in nucleic acids by means of a protein-synthesizing apparatus involving three different kinds of RNA.

In brief, the genetic material, through the process of transcription, produces a molecule of *messenger RNA* (*mRNA*) that is, base for base, a complement to the bases on one of the DNA strands. Through the mediation of ribosomes, which themselves consist of *ribosomal RNA* (*rRNA*) and protein, a sequence of bases in mRNA then translates into a sequence of amino acids. This translation follows the triplet rule that a sequence of three mRNA bases designates 1 of the 20 different kinds of amino acids used in protein synthesis.

Note that during the translation, no physical material is actually inserted by the mRNA into the protein; only information is transferred. That is, the mRNA only designates the linear position in which each amino acid is to be placed by special molecules of *transfer RNA* (*tRNA*) that bring the amino acids to the messenger. With the aid of the ribosome and various enzymes, the amino acids then connect in sequence through peptide linkages. Thus a polypeptide chain of amino acids forms in which the genetic material has ultimately designated the precise position of each component. DNA (or RNA in some viruses) thus provides the *genotype*, or genetic endowment of an organism. The expression of this nucleic acid information, via transcription and/or translation, provides the various aspects of an organism's appearance, or *phenotype*.

The presently observed circular interdependency of all these events, such as the replication of nucleotides because of the presence of appropriate enzymes and the determination of enzyme structure because of the presence of appropriate nucleotide sequences, points to certain difficulties in finding a reasonable explanation for the origin of life. Which of the many biochemical agents came first? How did they arise? How could they have functioned before an entire cellular structure formed? Many proposals for the origin of life exist, and we can generally divide such proposals into two broad categories: life on Earth developed from previous life, or life on Earth originated by chemical means.

Life only from Prior Life

The concept that life did not originally arise on Earth is embodied in almost all the creation myths of humans. These myths usually presume the special creation of life on Earth by one or more superior, intelligent, and all-powerful beings who themselves possess attributes of life such as sensation, thought, and purposive movement. This concept has the advantage that it explains the origin of terrestrial life in a simple fashion that is especially attractive to those who believe that natural events are governed by conscious agents. It does not explain the source of the initial creator, and therefore does not explain life's origin.

Another ancient concept is that life can arise spontaneously at any time, such as the presumed origin of insects from sweat and crocodiles from mud. Strangely enough, this view was often held simultaneously with the view that life derives from a conscious creator, and was popular in Europe throughout medieval times. Pasteur and others put spontaneous generation theory to rest in the 1860s, and it has not since been revived in its original form. As we shall see later, the modern concept of the spontaneous origin of life on Earth does not include such simple means as the immediate action of sunlight on liquid or clay but proposes instead the past existence of more complex yet more understandable biochemical processes.

One variation of the theory that life comes only from life is the proposal that life is somehow ingrained in all matter, and the creation of matter by whatever cause is responsible for the creation of life. Although this notion offers the advantage of ascribing life to a natural event, its origin would seem difficult if not impossible to understand. Obviously, many aspects of matter show no evidence of life if we define life to include those attributes possessed by terrestrial organisms, such as metabolism, reproduction, and so on. How did these attributes arise?

Another variation suggests that life on Earth arose elsewhere, perhaps on a distant planetary body circling a distant star, and was then transported to Earth by space-resistant spores or other means. This notion, called *panspermia*, was fostered by the chemist Arrhenius (1859-1927) in the early part of this century and still has some adherents among scientists today. It overcomes the difficulty of seeking a chemical explanation for the origin of life on Earth but does not, of course, explain the distant origin of life.

Proponents of the panspermia hypothesis point to the discovery of a number of different organic compounds in carbon-containing meteorites (*carbonaceous chondrites*), ranging from carbohydrates to amino acids. Looked at closely, however, the structural forms or isomers of amino acids in these carbonaceous chondrites possess the two different kinds of *optical activity* (dextro- and levorotary) in equal amounts, therefore comprising a *racemic mixture* that shows no optical activity. In contrast, the amino acids of living forms generally show optical activity of only one type, levorotary. Furthermore, a number of the amino acids in meteorites do not appear in proteins. Along with other observations, these findings indicate that organic compounds probably formed through random chemical reactions in the meteorite itself or in its parent body, rather than through ordered living processes.

Opposition to the panspermia hypothesis also notes the difficulty of envisaging how spores or "bugs" from outer space can get to Earth without the help of conscious agents in spaceships. If the bug is too large, it cannot be easily ejected from its home planet nor subsequently

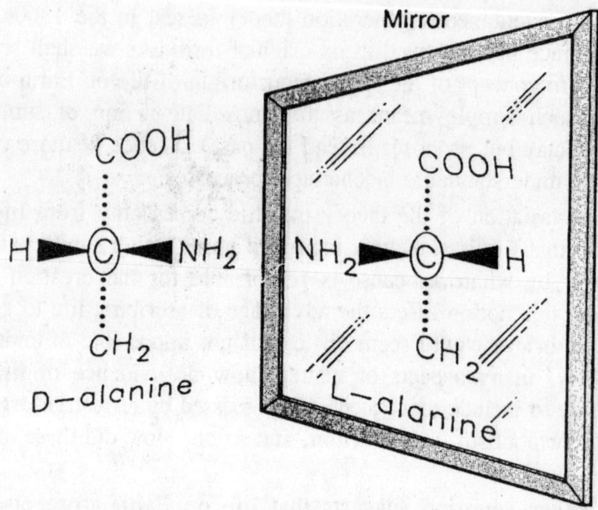

Fig. 2.5. Structures of an L-form and a D-form amino acid. The dark, wedge-shaped bonds indicate that the attached NH_2 and H groups project above the plane of the paper.

pushed out by sun radiation in its particular solar system. If it is too small, it will be kept from entering Earth's field by radiation pressure from our sun. Shklovskii and Sagan suggest that bugs larger than 0.6 μm cannot escape from Earth, while bugs smaller than 0.6 μm would be pushed away from Earth. So the donor planet must have been capable of ejecting bugs that we cannot eject (e.g., about 1 μm); that is, the sun of the donor planet must have been very hot (high radiation pressure). But if such a sun were hot enough for this purpose, it would destroy ejected particles by radiation.

In addition, the hazards of interstellar travel are many. For example, ultraviolet radiation (UV) from our sun will kill ejected particles from Earth within about one day of interplanetary travel. If the bugs were shielded from UV, they would be too heavy to be ejected. Among other space hazards are the hot, ionized gases that surround early-type stars, the presence of cosmic rays, and the absorption of bugs into passing suns by gravitational attraction.

However, even assuming that bugs survived all other obstacles, the vastness of space would disperse these spores so widely that their chance of reaching Earth would be infinitesimal. Shklovskii and Sagan have calculated that 100 million life-bearing planets in our galaxy would each have had to eject about 1,000 tons of spores in order for Earth to have received a single microorganism during its first billion years of history. Such difficulties make questionable whether an event as seemingly rare as panspermia is more probable than the chemical origin of life on this planet: Why should the origin of life on other bodies have had a greater probability than its origin on Earth? Such misgivings, along with a rapid increase in our understanding of molecular biology and biochemistry, have influenced most scientists to concentrate their attention on a terrestrial origin of life.

Terrestrial Origin of Life

The difficulty in visualizing life originating on Earth is essentially one of visualizing the molecular environment and events that occurred in a long-distant past. Unfortunately, we have as yet no certainties about the details of our molecular past, and we may never have such knowledge because molecular fossils are indeed sparse. At best, we can try to deduce the general nature of some of the original molecular events from present living structures and reactions and try to reconstruct such events experimentally under controlled conditions. However, before undertaking such a molecular review, it is important to consider the framework in which people usually pose the question of the origin of life from a terrestrial source. That is, we must attempt to deal with the problem of whether a highly complex, ordered phenomenon such as life could have arisen at all from the molecular chaos assumed to have existed during the Earth's early history.

Obviously, the probability for a modern, self-reproducing cell arising from complete disorder is embarrassingly small. To use an oft-quoted example, can a monkey, given even billions of years, produce the works of Shakespeare by randomly pressing the keys of a typewriter? Even if we restrict Shakespeare's writings to 1 million (10^6) alphabetical letters and limit the typewriter to 26 keys, the chance for such an event is $(1/26)^{10^6}$. This means that even if a monkey could type 1 million words a second, we could expect such an event to occur only once in $7 \times 10^{1,414,965}$ years!

By similar reasoning, the chances for most complex organic structures to arise spontaneously are infinitesimally small. Even a small enzymatic sequence of 100 amino acids would have only one chance in 20^{100} ($= 10^{130}$) to arise randomly, since there are 20 possible kinds of different amino acids for each position in the sequence. Thus, if we randomly generated a new 100-amino-acid-long sequence each second, we could expect such a given enzyme to appear only once in 4×10^{122} years! In terms of the volume necessary to generate all such possibilities, the difficulty appears just as immense: if an entire universe 10 billion light-years in diameter were densely packed with randomly produced polypeptides, each 100 amino acids long, the number of such molecules—10^{103}—would not equal their 10^{130} possibilities.

These arguments long seemed formidable and were further strengthened by the suggestion that nature itself would deteriorate any complex organization of matter even if such complexity were to arise accidentally. Theorists often pointed out that according to the *second law of thermodynamics*, the energy in a system tends toward diffusion rather than concentration; that is, *entropy* (disorder) increases rather than decreases. Thus, there appeared to be only a negative answer to the question that Pasteur had posed in the nineteenth century: "Can matter organize itself?" It seemed either that the living organization of matter must be explained as arising from a mystical nonnatural source, or that this event, if it were of natural origin, was so improbable that any attempt at comprehension or reconstruction would be meaningless.

In answer to these apparent difficulties, many scientists today point to two important considerations:

- The likelihood that primeval "living" organisms did not have many of their present complexities.
- The formation of organic molecules and subsequent organic structures was not the result of completely random events, although such events were nevertheless natural in the sense that they followed chemical and physical laws.

The first consideration, that early organisms were more primitive than those of today, is extremely important. Perhaps the most basic quality of life we would recognize in even a primitive living organism is its ability to perform those reactions necessary for it to grow and replicate, certainly the endless loop of metabolism and information transfer that we now see embodied in the intricate relationship between proteins and nucleic acids need not always have been of the same complexity. As we shall see later, protein-like compounds may arise in reaction mixtures of amino acids without intervention by nucleic acids and, although formed randomly, these compounds may then function in a variety of enzymatic ways.

Although it is true that chemically generated proteins do not have the repeatable and precisely ordered sequences of amino acids in cellular proteins, it is probably also true that the metabolic functions necessary for survival and growth were much simpler in the past. The relative simplicity of early "life" and its precursors, especially in the absence of competition with the more sophisticated later forms, seems a reasonable assumption to make.

The second consideration, that life did not arise from absolute chaos, has been supported in many ways. Many scientists have pointed out that the evolution of our solar system offered a number of essential prerequisites that enabled the development and sustenance of life:

1. Our planet possessed a sun of moderate size that was on the main sequence of stellar evolution. This provided a steady rate of emitted radiation over a long enough period of time for life to develop. The amount of energy (available from solar radiation seems to have always been far greater than any other source provided, although energy from other sources may also have been important in initiating particular chemical reactions.
2. A fairly large sampling of different elements existed, such as H, O, C, N, S, P, Ca, and others. These elements must have provided considerable chemical diversity, enabling reactions to occur that were necessary to form organic molecules involving carbon. The important chemical attributes of carbon, its ability to form four covalent bonds and the tetrahedral arrangement of its outer electrons, provided the opportunity for the formation of a large number of different kinds of stable molecules with considerable three-dimensional variety and complexity. Interestingly, the terrestrial presence of such molecules is not unique: by means of spectroscopy, researchers can now observe a variety of organic molecules in the dense interstellar clouds that give rise to stars and planets. Such observations indicate that a number of compounds necessary for the origin of life were already present both before and during the formation of our solar system, and their synthesis was *abiotic*-independent of living systems.
3. The Earth followed a nearly circular orbit at a fairly uniform distance from the sun. Such an even orbit would eliminate temperature extremes in which organic molecules would be incapable of forming or functioning.
4. There was present on Earth large amounts of an excellent solvent, water, which is stable in liquid form over a relatively wide range of temperatures and enables both acids and bases to ionize and react. Water also has the advantage that it floats in its crystalline frozen form (ice), so that bodies of water containing organic matter may remain liquid under a surface of ice, rather than freezing because of the subsurface accumulation of ice. Geochemists now

believe that water must have been present early in the Earth's history in the cold planetesimal condensations and appeared in liquid form as soon as the lithosphere reached appropriate temperatures. Additional water has been continually emitted into the atmosphere through volcanic activity, which was probably greater in the past than at present. Since water causes crustal erosion that leads to sedimentary rock formation, the presence of significant amounts of water dates back to the beginning of the geological record.

5. Hydrogen-containing gases existed for a long initial period in the Earth's history, derived from the high cosmic abundance of hydrogen and the consequent abundance of its compounds. Even though free hydrogen was probably lost early in the Earth's history, outgassing from the Earth's interior would have led to at least a partially reducing atmosphere in some or many localities in which hydrogen protons could be donated to a variety of proton-accepting elements, especially carbon. Such gases, along with the energy from solar and ultraviolet radiation, therefore enabled the formation of a variety of organic molecules that could, in turn, provide both structure and energy for living processes; for example, amino acids, sugars, fatty acids, purines, and pyrimidines.

In summary, we can say that there was considerable *molecular preadaptation* for the biochemical events leading to the formation of life even before life appeared. That is, there were appropriate energy sources, chemicals, temperature, and solvent—the foundations of a "universal organic chemistry." What kinds of reactions would then have taken place?

Origin of Basic Biological Molecules

In the 1920s, Oparin, a Russian biochemist, and Haldane, an English geneticist, independently suggested that the primitive atmosphere of the Earth was reducing and that organic compounds formed in such an atmosphere might be similar to those presently used by living organisms. However, almost 30 years elapsed before someone undertook an experimental test of this hypothesis. In 1953, Miller placed together in a glass apparatus methane, ammonia, and hydrogen gases. He generated an electric spark in a large 5-liter flask, and boiled water in a smaller flask to provide vapour to the spark as well as to circulate the gases. Compounds formed by sparking were then condensed, or recirculated if they were volatile. After one week of continuous electrical discharge, he chromatographed and analyzed the products

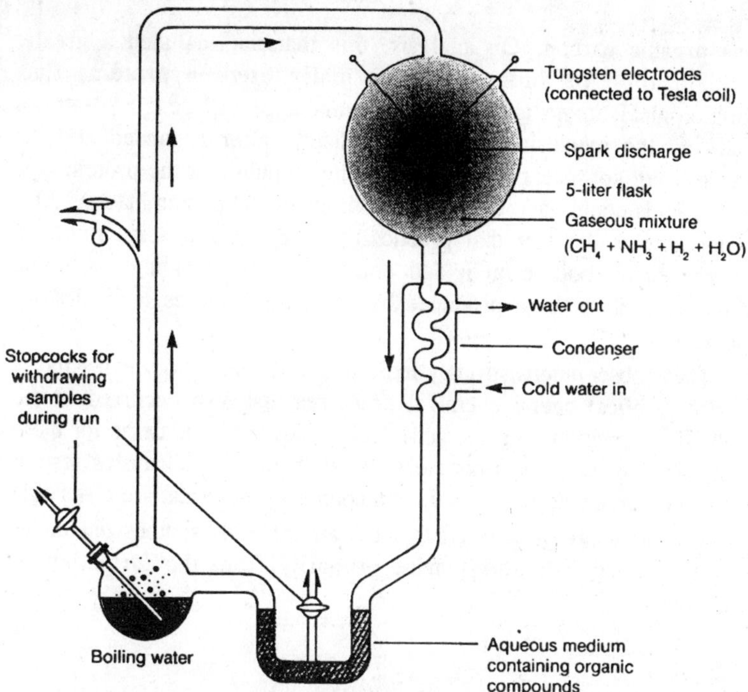

Fig. 2.6. Apparatus Miller (1953) used to demonstrate the synthesis of organic compounds by electrical discharge in a reducing atmosphere.

accumulated in the aqueous phase. Note that a large portion of these compounds are relatively simple and include both amino acids and other substances, such as urea, found in living organisms. In fact, of the wide array of possible complex molecules that such apparently random chemical reactions could have produced, it is remarkable that significant amounts of such relatively simple compounds essential to life actually formed. These experiments and others that followed therefore point strongly to the likelihood that the chemical environment that existed before the origin of life was probably not "chaos." Rather, the Earth had a significant amount of simple organic molecules that could participate in forming living organisms.

Moreover, astronomers can see such compounds in interstellar clouds in our galaxy and also in various carbonaceous meteorites that they believe represent material remaining in space from the original solar condensation 4.6 billion years ago. One such example, the Murchison meteorite that fell in Australia in 1969, contained more than 80 kilograms of carbonaceous material, of which about 1 percent was organic carbon. On analysis, this material included a greater concentration of amino acids normally used in proteins than nonbiological amino acids. Since a number of different laboratories analyzed the meteorite almost immediately after it landed and the results were consistent overall, researchers doubt that the protein-type amino acids could have originated from terrestrial contamination. Also remarkable is the fact that practically all the amino acids found in the meteorite, both protein and nonprotein, are identical to amino acids researchers have produced by sparking mixtures in laboratory experiments.

These observations strongly indicate that the laboratory experiments probably reflect actual chemical processes that also occurred in the synthesis of prebiotic organic compounds. However, the cause for these impressive correlations is not clear: What chemical mechanisms restrict or bias the production of organic compounds to those that are observed?

Theorists have suggested that the amino acids synthesized under primitive Earth conditions arise primarily from the formation of aldehydes,

$$R\text{—}\overset{\overset{\displaystyle O}{\|}}{C}\text{—}H$$

(where R may represent any group), which then interact with ammonia and cyanide compounds. These reactive chemicals may have arisen from a variety of simple gases or from their further interactions:

$$\left.\begin{array}{l} N_2 + H_2 \\ N_2 + H_2O \end{array}\right\} \longrightarrow NH_3 \text{ (ammonia)}$$

$$CH_4 \longrightarrow C_2H_2 \text{ (HC}\equiv\text{CH, acetylene)}$$

$$\left.\begin{array}{l} CH_4 + N_2 \\ CH_4 + NH_3 \\ CO + NH_3 \\ C_2H_2 + N_2 \end{array}\right\} \longrightarrow HCN \text{ (HC}\equiv\text{N, hydrogen cyanide)}$$

$$\left.\begin{array}{l} CH_4 + H_2O \\ CH_4 + CO_2 \\ CO_2 + H_2 \\ CO_2 + H_2O \end{array}\right\} \longrightarrow HCHO \text{ (H}\overset{\overset{O}{\|}}{C}\text{H, formaldehyde)}$$

$$CH_4 + H_2O \longrightarrow CH_3CHO \text{ (}H_3C\text{—}\overset{\overset{O}{\|}}{C}\text{H, acetaldehyde)}$$

According to one of the possible pathways of amino acid synthesis (the Strecker synthesis), subsequent steps are as follows:

$$R-\overset{\overset{O}{\|}}{C}-H + NH_3 \longrightarrow R-\underset{\underset{NH_2}{|}}{\overset{\overset{OH}{|}}{C}}-H \longrightarrow \underset{\text{Aldimine}}{R-\underset{\underset{NH}{\|}}{C}-H} + H_2O$$

$$R-\underset{\underset{NH}{\|}}{C}-H + HCN \longrightarrow \underset{\text{Aminonitrile}}{R-\underset{\underset{NH_2}{|}}{\overset{\overset{H}{|}}{C}}-C\equiv N}$$

$$R-\underset{\underset{NH_2}{|}}{\overset{\overset{H}{|}}{C}}-C\equiv N + H_2O \longrightarrow \underset{\text{Aminoamide}}{R-\underset{\underset{NH_2}{|}}{\overset{\overset{H}{|}}{C}}-\overset{\overset{O}{\|}}{C}-NH_2}$$

$$R-\underset{\underset{NH_2}{|}}{\overset{\overset{H}{|}}{C}}-\overset{\overset{O}{\|}}{C}-NH_2 + H_2O \longrightarrow \underset{\alpha\text{-amino acid}}{R-\underset{\underset{NH_2}{|}}{\overset{\overset{H}{|}}{C}}-\overset{\overset{O}{\|}}{C}-OH} + NH_3$$

If R in the preceding reactions is a hydrogen atom—that is, if the initial molecule,

$$R-\overset{\overset{O}{\|}}{C}-H$$

is formaldehyde (HCHO)—then the resulting amino acid is glycine. Glycine can also result from adding water (hydrolysis) to cyanide polymers:

$$\underset{\text{Cyanide monomers}}{H-C\equiv N + H-C\equiv N} \longrightarrow \underset{\text{Dimer}}{H-\overset{\overset{NH}{\|}}{C}-C\equiv N}$$

$$H-\overset{\overset{NH}{\|}}{C}-C\equiv N + H-C\equiv N \longrightarrow \underset{\text{Trimer}}{N\equiv C-\underset{\underset{NH_2}{|}}{\overset{\overset{H}{|}}{C}}-C\equiv N}$$

$$N\equiv C-\underset{\underset{NH_2}{|}}{\overset{\overset{H}{|}}{C}}-C\equiv N + H-C\equiv N \longrightarrow \underset{\text{Tetramer}}{(H_2N)(N\equiv C)C=C(H_2N)(C\equiv N)}$$

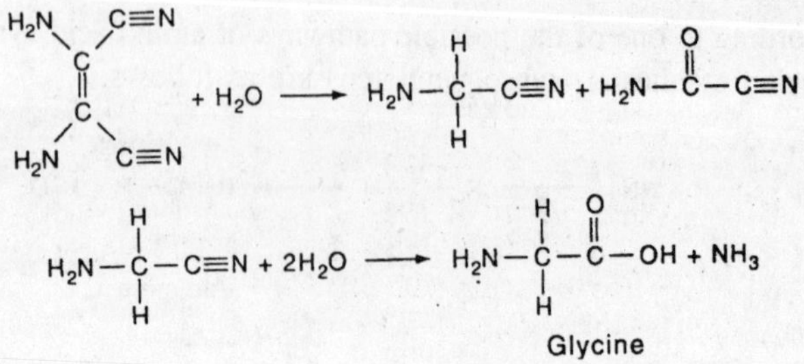

$$(H_2N)C(C\equiv N)=C(C\equiv N)(NH_2) + H_2O \longrightarrow H_2N-CH_2-C\equiv N + H_2N-C(=O)-C\equiv N$$

$$H_2N-CH_2-C\equiv N + 2H_2O \longrightarrow H_2N-CH_2-C(=O)-OH + NH_3$$

Glycine

Adding formaldehyde to glycine under alkaline conditions can then produce serine:

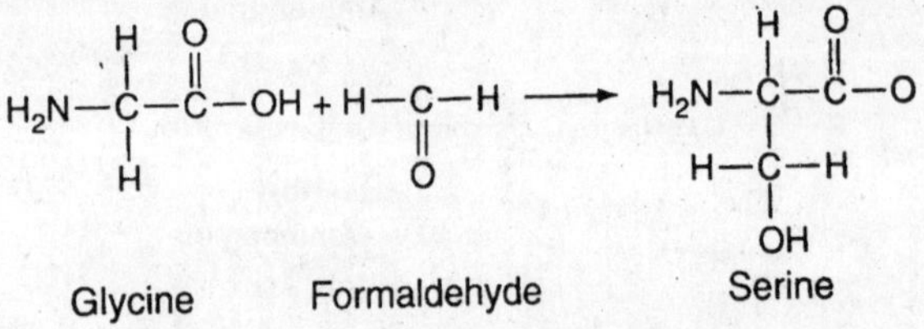

$$H_2N-CH_2-C(=O)-OH + H-C(=O)-H \longrightarrow H_2N-CH(CH_2OH)-C(=O)-O$$

Glycine Formaldehyde Serine

All 20 different amino acids now used in protein synthesis have a similar structural pattern, although many of them are synthesized by different biochemical pathways. Among other basic organic molecules

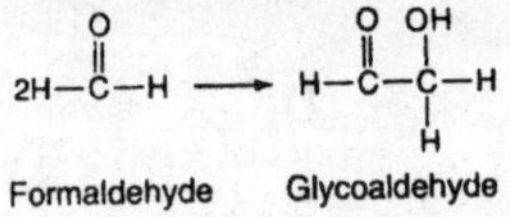

$$2H-C(=O)-H \longrightarrow H-C(=O)-CH(OH)-H$$

Formaldehyde Glycoaldehyde

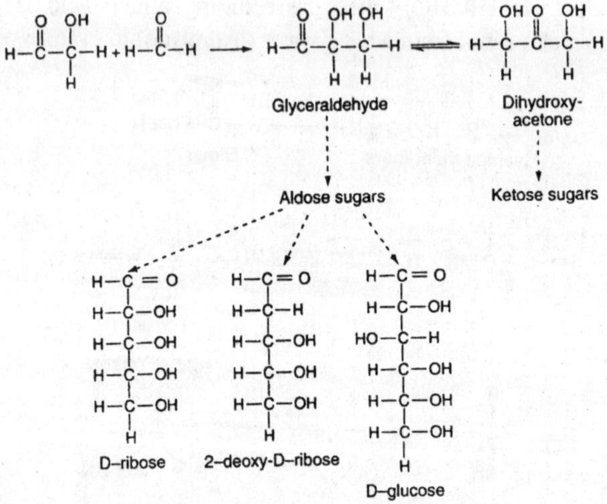

$$H-C(=O)-CH(OH)-H + H-C(=O)-H \longrightarrow H-C(=O)-CH(OH)-CH(OH)-H \rightleftharpoons H-CH(OH)-C(=O)-CH(OH)-H$$

Glyceraldehyde → Aldose sugars

Dihydroxyacetone → Ketose sugars

D-ribose: H–C=O, H–C–OH, H–C–OH, H–C–OH, H–C–OH, H

2-deoxy-D-ribose: H–C=O, H–C–H, H–C–OH, H–C–OH, H–C–OH, H

D-glucose: H–C=O, H–C–OH, HO–C–H, H–C–OH, H–C–OH, H–C–OH, H

that could easily be synthesized under fairly simple conditions are the *sugars*. Significant yields of glucose, ribose, and deoxyribose, for example, occur from condensing formaldehyde.

The *purine* and *pyrimidine* bases that are essential components of nucleic acids can also be synthesized under prebiotic conditions. For example, Oro and co-workers have shown that heating aqueous solutions of ammonium cyanide (prepared by the reaction of HCN with NH_4OH) produces up to 0.5 percent yield of adenine. Similarly, ultraviolet radiation acting on hydrogen cyanide solution produces a number of purines, including adenine and guanine. Condensation reactions in forming adenine have been studied in some detail, and researchers have suggested that one sequence may be as follows:

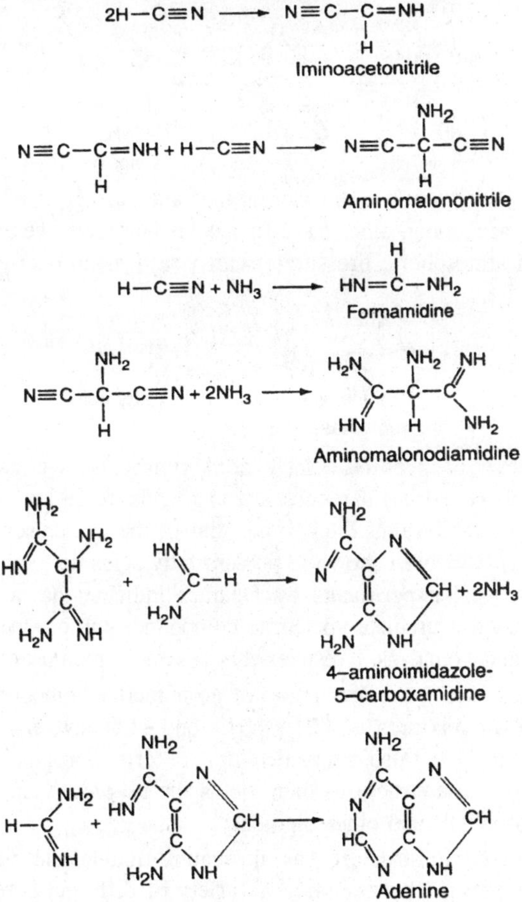

$$2H{-}C{\equiv}N \longrightarrow N{\equiv}C{-}CH{=}NH$$

Iminoacetonitrile

$$N{\equiv}C{-}CH{=}NH + H{-}C{\equiv}N \longrightarrow N{\equiv}C{-}CH(NH_2){-}C{\equiv}N$$

Aminomalononitrile

$$H{-}C{\equiv}N + NH_3 \longrightarrow HN{=}CH{-}NH_2$$

Formamidine

$$N{\equiv}C{-}CH(NH_2){-}C{\equiv}N + 2NH_3 \longrightarrow (H_2N)(HN{=})C{-}CH(NH_2){-}C({=}NH)(NH_2)$$

Aminomalonodiamidine

$$\text{Aminomalonodiamidine} + HN{=}CH{-}NH_2 \longrightarrow \text{4-aminoimidazole-5-carboxamidine} + 2NH_3$$

4–aminoimidazole-
5–carboxamidine

$$H{-}C(NH_2){=}NH + \text{4-aminoimidazole-5-carboxamidine} \longrightarrow \text{Adenine}$$

Adenine

The reaction of cyanoacetylene with cyanates such as urea has produced the pyrimidine cytosine as shown next, and researchers have proposed similar synthetic procedures for the other pyrimidines uracil and thymine:

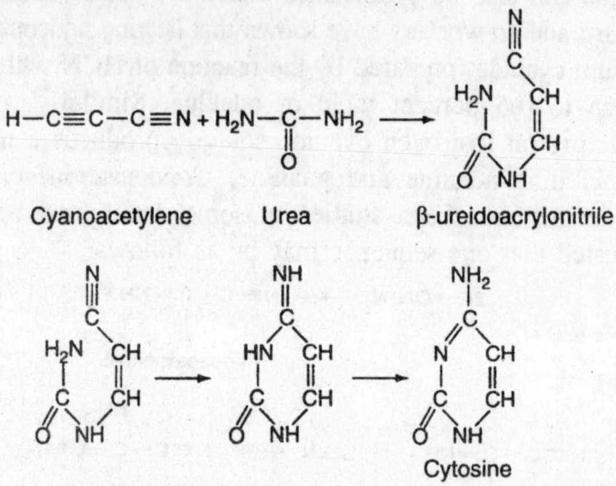

Fatty acids, now used in membranes and storage tissues of living organisms, are among other basic molecules that have been synthesized under high atmospheric pressures, with γ-rays as an energy source:

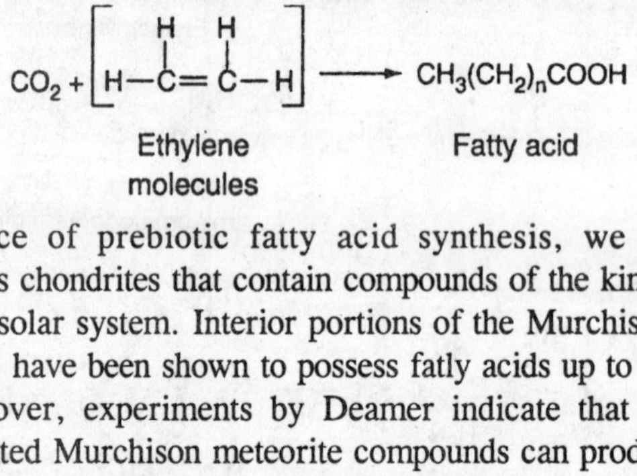

For evidence of prebiotic fatty acid synthesis, we can look to carbonaceous chondrites that contain compounds of the kind synthesized in the early solar system. Interior portions of the Murchison meteorite, for example, have been shown to possess fatly acids up to eight carbons long. Moreover, experiments by Deamer indicate that a portion of uncontaminated Murchison meteorite compounds can produce fatty-like structures and boundaried vesicles that resemble membranes.

Pyrroles, which are precursors of porphinelike compounds, can be synthesized in mixtures of CH_4, NH_3, and H_2O and can then react with formaldehyde (also benzaldehyde) to form porphine structures. Oxidizing of these structures then yields the *porphyrin* rings found in heme, chlorophyll, and other pigments.

The porphyrin structure has alternating double and single bonds that can "resonate" by assuming a variety of different configurations

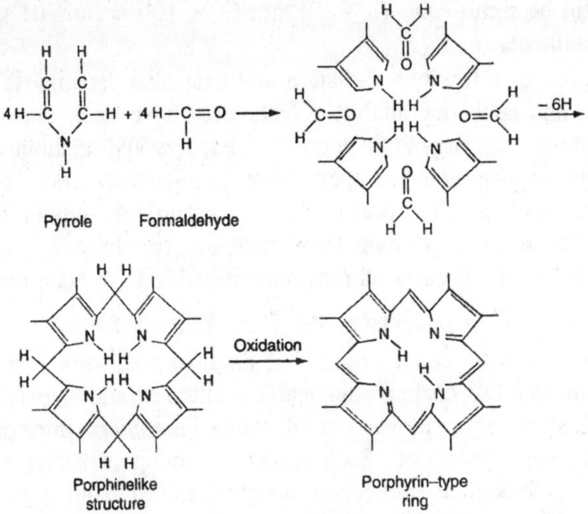

Pyrrole Formaldehyde

Porphinelike structure

Oxidation

Porphyrin–type ring

without changing the position of their constituent atoms. Such resonance confers stability on porphyrins, enabling them to hold extra electrons and thus to function as electron acceptors (oxidation) or electron donors (reduction). Similar oxidative and reductive functions can be performed by nucleotide derivatives such as nicotinamide adenine dinucleotide (NAD), called *coenzymes* because they act in union with protein enzymes to catalyze a wide variety of chemical reactions.

We can thus see that many of the basic organic molecules used in living organisms form relatively easily in many reactions. The amounts per reaction are usually small, but the overall quantities of such substances may have been quite large. Shklovskii and Sagan, for example, point out that 1 photon of ultraviolet radiation produces a quantum yield of about 1/100,000 to 1/1,000,000 of a simple organic molecule. If we take 10^{-22} grams as the average mass of such a molecule, then the quantum yield per photon is about $10^{-5} \times 10^{-22} = 10^{-27}$ grams.

Shklovskii and Sagan estimate the number of photons at the top of the Earth's atmosphere in primitive times at 3×10^{14} photons/cm^2/sec, so the quantum yield per square centimeter per second may have been $10^{-27} \times (3 \times 10^{14}) = 3 \times 10^{-13}$ grams. Thus, if the reducing atmosphere lasted for 300 million years (about 10^{16} seconds), enough energy would have formed to produce $(3 \times 10^{-13}) \times 10^{16} = 3 \times 10^{3}$ grams of matter per square centimeter of the Earth's surface. Furthermore, even if this material were diluted in as deep an ocean as the present (3×10^{5} centimeters), the concentration of the solution would still be significant: (3×10^3)gm/(3×10^5) cc, or .01 gram per cubic centimeter.

Of course, ultraviolet radiation and heat also decompose organic material, and such degradative effects may have been considerable. Nevertheless, once organic material formed, it would undoubtedly have had many opportunities to accumulate in relatively cool, protected localities, such as the fissures of rocks and the depths of pools inaccessible to decomposition by ultraviolet rays. In such places, the concentrations of organic materials may therefore have been quite high.

Condensation and Polymerization

Given localized concentrations of amino acids, sugars, and other organic molecules, further chemical evolution would depend on the polymerization or condensation of these *monomers* into peptides, polysaccharides, and so on. Such events are not spontaneous: to obtain one small polypeptide (molecular weight 12,000) from a one molar aqueous amino acid solution, in the absence of any other chemical forces, would require a volume of amino acids 10^{50} times that of the Earth. How then could such polymerizations occur?

Most polymerizations depend on the removal of water molecules from the monomers to be condensed. Peptide bonds, for example, ordinarily form in the cell on ribosomes through *phosphate bond* energy. Outside the cell, the task is more difficult, but bonds can nevertheless form either in aqueous medium or under anhydrous conditions.

So far, compounds identified as condensing agents that could have existed early in the Earth's history include the following:

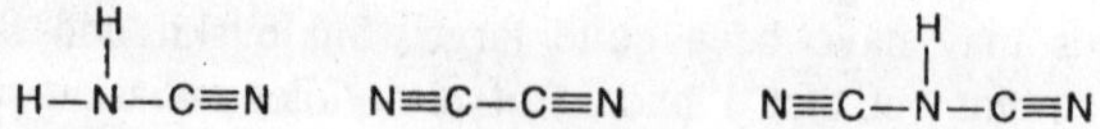

$$\mathrm{H{-}N(H){-}C{\equiv}N} \qquad \mathrm{N{\equiv}C{-}C{\equiv}N} \qquad \mathrm{N{\equiv}C{-}N(H){-}C{\equiv}N}$$

Cyanamide   Cyanogen   Dicyanamide

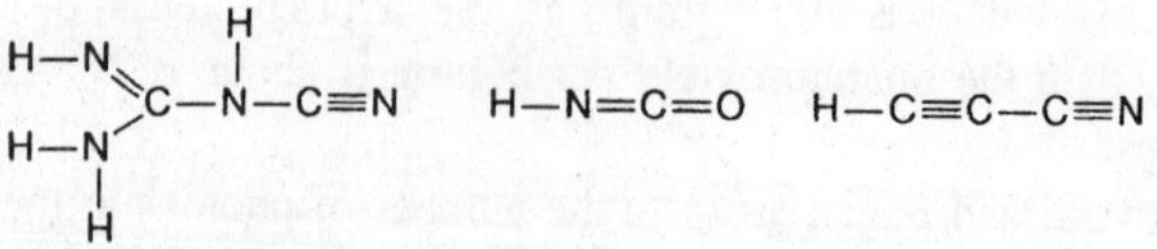

$$\mathrm{(H{-}N{=})(H{-}N(H){-})C{-}N(H){-}C{\equiv}N} \qquad \mathrm{H{-}N{=}C{=}O} \qquad \mathrm{H{-}C{\equiv}C{-}C{\equiv}N}$$

Dicyandiamide   Cyanic acid   Cyanoacetylene

In each case, the unsaturated cyano-carbon-nitrogen bonds enable the condensing agent to combine with water and release energy during this hydration. For example,

$$\underset{\text{Cyanamide}}{\mathrm{H{-}N(H){-}C{\equiv}N}} + H_2O \longrightarrow \underset{\text{Urea}}{\mathrm{H{-}N(H){-}C({=}O){-}N(H){-}H}} + \text{free energy}$$

Thus, the condensation of two amino acid into a dipeptide can couple to the *hydrolysis* of cyanamide:

$$\underset{\text{Amino acid 1 }(R_1)}{H-N(H)-C(H)(R_1)-C(=O)-OH} + \underset{\text{Amino acid 2 }(R_2)}{H-N(H)-C(H)(R_2)-C(=O)-OH} + \underset{\text{Cyanamide}}{H-N(H)-C\equiv N} \longrightarrow$$

$$\underset{R_1 - R_2 \text{ dipeptide}}{H-N(H)-C(H)(R_1)-C(=O)-N(H)-C(H)(R_2)-C(=O)-OH} + \underset{\text{Urea}}{H-N(H)-C(=O)-N(H)-H}$$

Because of their apparent preference for reacting with organic molecules carrying anions (e.g., phosphate HPO_4^-), many of the cyanic condensing agents produce peptide bonds between amino acids even in aqueous solutions. Some, such as cyanogen and cyanamides, also cause nucleotides to form by the phosphorylation of adenosine, uridine, and cytosine; for example,

$$\underset{\text{Adenosine}}{(NH_2)\text{-base-}O\text{-ring-}CH_2OH} + \underset{\text{Orthophosphate}}{H_3PO_4} \xrightarrow{\text{Dicyan diamide}} \underset{\text{Adenylic acid (adenosine monophosphate or AMP)}}{(NH_2)\text{-base-}O\text{-ring-}CH_2OH_2PO_3}$$

Under anhydrous conditions, with no or few water molecules, heat can promote condensation and polymerization by causing the loss and evaporation of water molecules even in the absence of specific condensing agents. One such reaction, accomplished by heat in the laboratory and by enzymes in living organisms, is the formation of high-energy phosphate bonds from orthophosphate:

$$2\underset{\text{Orthophosphate}}{\left[O^- - P(=O)(OH) - OH\right]} \longrightarrow \underset{\text{Pyrophosphate}}{O^- - P(=O)(OH) - O - P(=O)(OH) - O^-} + H_2O$$

High yields of pyrophosphate can also be synthesized by the condensing agent cyanic acid (cyanate) reacting on precipitated hydroxyapatite $[Ca_{10}(PO_4)_6\,(OH)_2]$, a major phosphate mineral. Such

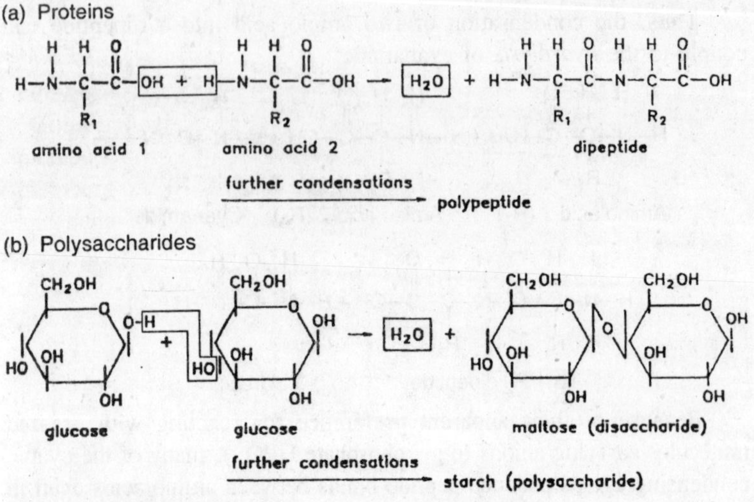

(a) Proteins

amino acid 1 + amino acid 2 → H_2O + dipeptide

further condensations → polypeptide

(b) Polysaccharides

glucose + glucose → H_2O + maltose (disaccharide)

further condensations → starch (polysaccharide)

Fig. 2.7. Examples of condensation reaction leading to the formation of peptides and polysaccharides.

pyrophosphates can then be made available for forming adenosine diphosphate (ADP) and *adenosine triphosphate* (*ATP*), reactions that can then be reversed by hydrolysis to yield energy:

$$\text{ATP} + H_2O \rightarrow \text{ADP} + \text{orthophosphate} + 7.50 \text{ kilocalories per molecular weight}$$

$$\text{ADP} + H_2O \rightarrow \text{AMP} + \text{orthophosphate} + 7.50 \text{ kcal/mole}$$

$$\text{AMP} + H_2O \rightarrow \text{adenosine} + \text{orthophosphate} + 3.40 \text{ kcal/mole}$$

When one or more steps in this hydrolytic sequence occurs, the cell gets its main source of energy and the primary means of removing further water molecules during condensation reactions. Biochemists have, in fact, suggested that polyphosphate chains may have provided some of the first organismic energy sources, and the adenosine component in ATP added later to act as a label that would allow enzymatic recognition.

Proteinoids

In the 19505, Fox and co-workers developed a technique in which heat could also be used to produce peptides from dry mixtures of amino acids. Depending on the kinds of amino acids in the mixture, they found that temperatures of 150° to 180°C could produce as much as 40 percent yield of peptidelike products with molecular weights between 4,000 and 10,000 daltons. Fox called these polymers *proteinoids* (also *thermal proteins*), and he and his group have shown that these compounds bear remarkable proteinlike features. According to their analyses, the proteinoids possess nonrandom proportions of amino acids; that is, their compositions are not simply based on the frequency of the different amino acids in the initial mixture.

They also suggest that the positions of the amino acids in the polymer are not based on their overall frequencies in the chain, since some amino acids preferentially occupy the N- and C-terminals of the proteinoids. The nonrandomness of proteinoid structure also seems supported by the fact that these polymers all show similar properties as tested by sedimentation rates, electrophoretic techniques, column fractionation, and other measurements. Thus, some preferential interaction between amino acids in proteinoid formation seems to dictate their position and frequency and lead to some degree of uniformity in the kinds of molecules produced.

Although not all the amino acid bonds formed in such proteinoids are of the peptide variety, nor do the shapes of these molecules follow the familiar α-helix of protein structure, there still seem to be enough peptide linkages to characterize them as proteins in many tests. Thus, proteinoids give positive colour tests with the same reagents that proteins do; their solubilities resemble proteins; they are precipitable with similar reagents and have other proteinlike traits.

Most importantly, proteinoids engage in a number of enzymelike activities that can increase the rates of various organic reactions. For example, they help split apart certain molecules by addition of water (hydrolysis), they catalyze the condensation of nucleotides such as ATP into di- and trinucleotides, and they help to remove carboxyl groups or amino groups from various structures. Moreover, they can improve catalytic activity of molecules, such as heme, that aid hydrogen peroxide in removing hydrogen from reduced compounds in oxidation reactions.

Fox and Dose have suggested that some of these reactions, combined into a particular sequence, may have served as the beginnings of later metabolic systems. Thus, decarboxylation of oxaloacetic acid can be followed by decarboxylation of its product, pyruvic acid, leading to acetic acid and carbon dioxide; or amination of pyruvic acid can lead to alanine. Furthermore, some proteinoids even show relatively sophisticated hormonal activity and can stimulate the production of melanin-producing cells.

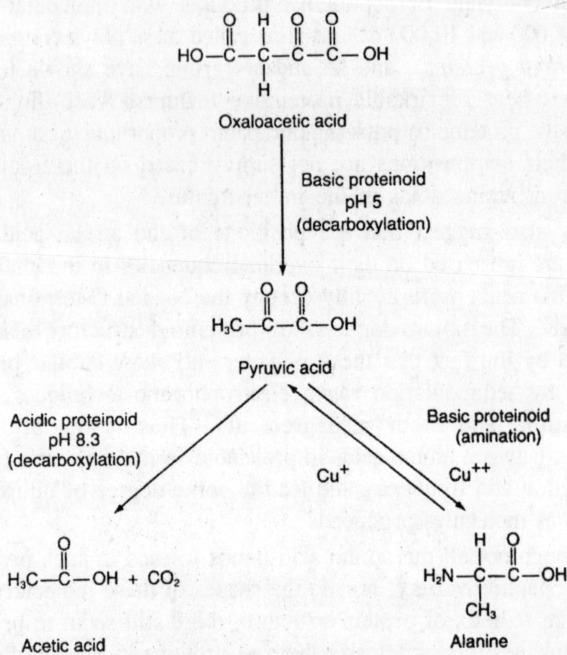

Fig. 2.8. Some sequential reactions catalyzed by different proteinoids or proteinoid complexes.

Although researchers have debated whether the thermal synthesis of proteins could occur extensively in present natural surroundings, the exact conditions encountered on the primitive Earth are certainly not known. Surfaces near some volcanic regions may have maintained appropriate temperatures for the condensation of amino acids, and cooling rains may have dispersed such thermally produced proteinoids to places where further interactions could lake place.

In any case, a wide-enough array of condensation mechanisms have been established, one or more of which most probably occurred in the past. Paecht-Horowitz and co-workers, for example, have shown that phosphate-activated amino acids such as aminoacyl adenylates will condense to form high yields of polypeptide chains on layered clays such as montmorillonite:

$$n\left[\mathrm{H{-}N(H){-}C(H)(R){-}C({=}O){-}O{-}P({=}O)(OH){-}O{-}adenosine}\right] \xrightarrow{\text{clay}}$$

Aminoacyl adenylate

```
     H  H  O      H  H  O
     |  |  ||     |  |  ||
H—N—C—C—O—N—C—C—O—· · · + nAMP
        |            |
        R1           R2
```

Polypeptide ($R_1 - R_2 \cdots R_n$)

The amino acid ends of the adenylates apparently penetrate the narrow layers of the clay, and the condensation reactions take place there. Some clays may also polymerize nucleotides, and Burton and co-workers report that certain nucleotide compounds that are strongly absorbed by montmorillonite clays form small amounts of dinucleotides. Thus, the early availability of cyanamides, heat, clays, and other condensing agents make it highly probable that polypeptides, polysaccharides, lipids, and perhaps even polynucleotides were present early in the Earth's history, and could have been used for primitive organismlike reactions and structures.

Origin of Organized Structures

The presence of appropriate organic monomers and polymers is only a first step in the origin of life. Living processes of metabolism and function occur because the materials of which organisms consist are highly organized. How did such organization come about?

At its earliest, interactions between molecules must have led them to assume relative positions based on forces such as hydrogen bonding, ionization, solubility, adhesion, surface tension, and so on. Phospholipids, for example, are organic molecules with a phosphorus-containing polar group at one end and nonpolar fatty acid groups at the other end. In water, a polar solvent, the polar ends of these molecules are oriented toward water (hydrophilic), while their nonpolar ends are oriented

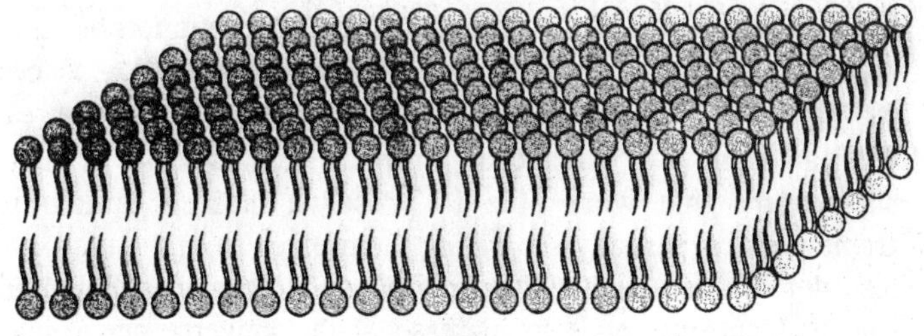

Bimolecular layer

Fig. 2.9. A diagrammatic view of a bimolecular sheetlike double layer of phospholipid molecules that have self-assembled with their hydrophilic phosphate heads facing the water solvent, and their hydrophobic hydrocarbon tails facing each other.

toward each other, away from water (hydrophobic). As a result, phospholipid membranous structures can form quickly yielding *vesicles* composed of bimolecular layers in which the nonpolar surfaces of each of the two layers "dissolve" in each other. Carried further, such vesicles can encapsulate inclusions in tide pools that undergo drying and wetting cycles.

Membranous droplets or vesicles composed of lipids, polypeptides, or other molecules undoubtedly formed in great quantities, produced by the mechanical agitation of molecular films on liquid surfaces or even spontaneously. The attainment of such droplet levels of organization was an important step in the origin of life for a number of reasons:

1. Depending on its structure and permeability, the membrane surrounding the droplet can selectively choose which compounds can enter from the environment and exit from the droplet.
2. Such *selective permeability* allows concentrations of particular compounds to differ across the membrane, enabling reactions to occur within the droplet that would not have occurred outside the droplet.
3. The presence of a basic protein causes a 100-fold increase in the entrapment of nucleic acids into such droplets. As Jay and Gilbert point out, "Protein-mediated encapsulation creates high local concentrations of protein and nucleic acids within the vesicular volume. . . . This would enhance the interaction of molecules with low affinities, potentiating the formation of aggregates with biological function"
4. The small size of the droplet can permit a chain or network of reactions to occur, the products of one reaction being available to serve as the substrates for another reaction.
5. Both the small size of the droplet and the concentration of various materials within it would permit localized precipitation to occur as well as the organization of compartments and substructures.
6. We can think of "advanced" droplets of this kind as unique subsystems that were able to preserve their organizational framework by partially separating themselves from the entropy or disorder in their surrounding environment. That is, although it is true that entropy tends to increase in the universe according to the second law of thermodynamics, it can nevertheless decrease in such subsystems during their life spans. Because of their semipermeable membranes they can use entering energy and matter to retain, and even enhance, their organizational and informational

structures as long as they can continue to perform biological processes.

It seems presumptuous and unrealistic to assume that the only kinds of organization capable of growth, metabolism, and reproduction are the structures found in present-day organisms. Although present forms are highly efficient, early forms could have functioned at a much lower level of efficiency, because they were not then competing with the more advanced forms. For a primitive form to show some (but certainly not all) "living" attributes, it would have been sufficient if it could merely grow (e.g., increase in size), maintain its individuality, and divide. It is therefore interesting to note that some authors ascribe such properties to bimolecular vesicles, and there are, in addition, at least two types of fairly simple laboratory-produced structures that seem to possess some aspects of these basic prerequisites: Oparin's coacervates and Fox's microspheres.

Coacervates

Coacervates have long been known to occur when dispersed colloidal particles separate spontaneously out of solution into droplets because of special conditions of acidity, temperature, and so on. If there is more than one type of macromolecular particle in the colloid, complex coacervates can form that show a number of interesting properties:

- They possess a simple but persistent organization.
- Although they are mostly unstable, some coacervates can maintain themselves in solution for extended periods.
- They can increase in size.

Oparin, the first to draw serious attention to these droplets, developed artificial coacervate systems that could incorporate enzymes that performed functions such as the synthesis and hydrolysis of starch

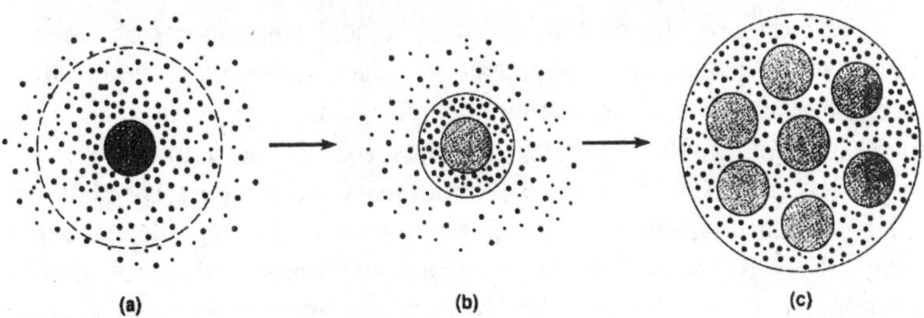

Fig. 2.10. Formation of coacervates by the exclusion of water molecules from associated colloidal particles.

as well as synthesizing polynucleotides. In a coacervate system containing chlorophyll irradiated with visible light, Oparin and co-workers showed that there can be a constant inflow of reduced ascorbic acid and oxidized methylene red, which then converts into a constant outflow of oxidized ascorbic acid and reduced methylene red. The chlorophyll picks up electrons from the ascorbic acid and then supplies those for the methylene red reduction—a process similar to common noncyclic photosynthesis in which water molecules supply electrons for reducing the coenzyme $NADP^+$ to NADPH.

Microspheres

Fox showed that these small spheres formed when the thermally produced proteinoids were boiled in water and allowed to cool. The microspheres are uniform in size, stable, bounded by double membranes that appear somewhat cell-like, and can undergo fission and budding. They appear in large numbers: 1 gram of proteinoid material can produce 10^8 or 10^9 microspheres. Among microsphere qualities indicating active internal processes are their selective absorption and diffusion of certain chemicals but not others, their growth in size and mass, and observations demonstrating osmosis, movement, and rotation. Moreover, microspheres show the potential for transferring information, in that proteinoid particles pass through junctions between them.

The spontaneous *self-assembly* of macromolecules into vesicles, coacervates, and microspheres indicates that the occurrence of similar entities under primitive conditions would probably not have been an unusual event. Such entities are not cells, of course, and considerable time may well have elapsed before more elegant structures with more complex metabolic capabilities could develop. Nevertheless, considerable evidence shows that the component materials of even more complex structures can self-assemble without the immediate presence of a prior pattern.

For example, the protein and nucleic acid components of tobacco mosaic virus spontaneously aggregate into the exact configuration needed to produce an active virus. A more complex virus such as T4, containing many different kinds of protein, also has a significant number of steps in which self-assembly occurs. Nomura and co-workers have shown even cellular organelles such as ribosomes to be capable of being formed by self-assembly from component materials. Each level of self-assembly, from monomers to polymers to coacervates to various cellular organelles, may thus have derived from nonrandom events, in that certain combinations form more quickly and easily than others.

Nonrandom self-assembly, however, is hardly sufficient to account for more than a few complexities of life. In their most essential aspects, such as the precisely ordered monomer sequences found in proteins and nucleic acids, present forms of life did certainly not result from mere chemical attractions between component amino acids or nucleotides. At the same time, we also have good reason to believe that, because of their specificity, the positioning of monomers in these precisely ordered sequences may nevertheless have had a nonrandom basis. To recapitulate a theme raised previously: How can the nonrandom biological order of amino acid and nucleotide sequences arise from disorder? The answer lies in selection.

Origin of Selection

Primitive structures, whether coacervates, microspheres, or other localized organizations, would have had one important evolutionary feature: they would serve as the first distinctive, multichemical *individuals*, or *protocells*, that could interact as units with their environment. Together with their various neighbours and progenies, such individuals would form a group or *population* on which selection could act. That is, protocells incorporating those organizations and metabolic activities most successful in growth and division would increase most in relative frequency or in area occupied.

We can thus say selection arises when the following conditions are reached:

- A population of individuals exists.
- The properties of these individuals are governed by reactions in which they absorb and transform environmental material into their own material.
- Individuals differ in the efficiency with which these processes take place.
- Availability of materials and energy is limited so that not all types of individuals can form, nor can all types of individuals formed survive.

The mechanism that enabled the formation of protocells is a crucial issue in understanding selection itself. If protocells could have originated only by self-replication, this would indicate that selection had always operated on the same efficient basis as it does now, with advantageous traits rapidly transmitted to succeeding generations by fairly exact replicative mechanisms. However, if protocells initially formed only through acts of prevailing environmental chemistry, as many biologists now believe, then selection was not very efficient in the past and would itself have undergone evolution from chemical nonreproductive selection to the more modern biological natural selection.

That is, early selection would have been confined to the survival of nonreproductive individuals who could wrest the most material from their environment and transform it for their own benefit with the least expenditure of energy. Although differences among such individuals could not be precisely transmitted, the fact that some such individuals survived and others did not would undoubtedly have affected the composition and further interactions of succeeding groups.

Inheritance would therefore, at first, have been mostly a matter of transmitting molecular "things" that permit survival, rather than transmitting the nucleic acid patterns that produce the "things." The earliest forms of "living" individuals may well have replicated themselves poorly, yet passed on some of their metabolic and enzymatic properties, which continued to be selected and improved.

From a materialistic point of view, unless we postulate an accident of immense proportions and infinitely low probability, selection must have bridged the gap between chemical evolution (changes in the composition of nonreproductive or poorly reproductive molecules, coacervates, microspheres, and so on) and biological evolution (changes in inherited differences among reproductive organisms). So far, selection is the only natural mechanism we know that can account for the creative changes among nonreproductive individuals that could have led them in the direction of living organisms. Although the events may be complex, the device is simple: organisms that react to their environment with improved functional information replace those that lack such information.

Nevertheless, selection is not merely a passive agent that sifts the good from the bad, the adaptive from the nonadaptive, but, because of its historical continuity, enables a succession of adaptations to accumulate that lead to something entirely new. Selection thus acts as a creative force that has made possible biological organizations that would otherwise have been highly improbable. To use a previous example, a polypeptide chain consisting of a specific sequence of 100 amino acids has an extremely low probability (10^{-130}) of occurring spontaneously without selection. However, if each step in the growth of the chain attains a selective advantage when the correct amino acid inserts, then the probability of achieving a functionally advantageous polypeptide is almost immeasurably increased.

Once the game of life has begun, the evolutionary replacement of the players, whether they are genes, organisms, races, species, or other entities, becomes inextricably bound to their ability to play the game further—an ability that selection has previously molded and is now in turn measured anew by selection. Some random replacement of the players certainly occurs by accident rather than by selection, but participation in life is selective by its very nature since the resources of life are always limited in one way or another. Thus, it is not merely their origin by selection that characterizes living systems but their continued ability to subject themselves to selection. This ability has led to a coupling between function and reproduction that provides living forms with their relatively rapid evolutionary rates.

3

Theories of Organic Evolution

How old is the earth? When people asked that question early in the nineteenth century, most of the answers were probably in the range of five to ten thousand years. The most common belief at that time was that the earth and all living things on it had been created at one time and that they have remained essentially as they were created. By the end of that century, however, many scientists were convinced that there is good evidence to conclude that the earth is many millions of years old.

It had been suggested as early as the days of ancient Greece that modern living things might have arisen through *evolution*, a process of change over a long period of time. By the early 1800s a few scientists such as Buffon, Lamarck, and Diderot supported the idea that an evolutionary process had occurred. But there was the problem of time. It did not seem that the earth was old enough for such changes in living things to have occurred. Then geological discoveries made it clear that the earth was very much older than previously thought. These discoveries were to have a tremendous impact on biology.

With the new ideas about the age of the earth, it became possible to think seriously about the idea of evolutionary changes in living things over a very long period of time. These ideas were not widely accepted, however, until Charles Darwin provided the foundation upon which the modern *theory of organic evolution* has been built. In 1859 Darwin published *On the Origin of Species*, a thoughtful explanation of his views on the evolution of life on earth. In the more than 100 years that have elapsed since that time, the theory of organic evolution has gained almost unanimous acceptance among scientists and has been greatly strengthened by evidence obtained from a wide variety of scientific disciplines.

Charles Darwin and the *Beagle* Voyage

Charles Darwin was born in England into a wealthy family. His father was a prominent physician, and his mother was a Wedgwood, of the renowned china manufacturing family. Darwin himself married a Wedgwood, his cousin Emma Wedgwood. All his life Darwin loved nature. As a young man, he was an ardent naturalist; he roamed the fields and woodlands of the English countryside, studying plants and animals and observing rock formations. This interest continued into his college years, for although his father wanted him to become a physician, Darwin preferred to associate with professors of biology and geology. When it became obvious that Darwin would not become a doctor, his father attempted to guide him into a career in the clergy.. That profession, however, held little more appeal for him than did medicine.

Finally, in 1831, Darwin was offered the opportunity to travel as a naturalist on H.M.S. Beagle, a ship commissioned to sail around the world on a survey mission. Darwin saw in this post an opportunity to see more of the natural world than was afforded most people of his time. In December of 1831 Darwin embarked on a voyage that was to change the history of biological thought.

Lyell, Hutton, and Uniformitarianism

Darwin was not a good sailor; he spent much of his time in his bunk suffering from seasickness. During the early part of the voyage, he occupied his time by reading the first volume of Charles Lyell's *Principles of Geology*, in which Lyell further developed James Hutton's theory of *uniformitarianism* in geology. Hutton contended that the earth is not a static, unchanging sphere but that the present features of the earth's surface are the result of a continuous cycle of erosion and uplift. Hutton had seen evidence that the earth's continents have been worn away continually by the agents of erosion and that weathered rock debris is transported by rivers to the oceans, where the loose sediments are deposited in thick layers that are eventually converted into sedimentary rocks. Hutton also proposed that "subterranean forces" operate to uplift material from below sea level to form new land surfaces. Hutton's concept of uniformitarianism stated that these forces have acted at the same rate throughout the earth's history to produce similar geological events. (Hutton and Lyell's general ideas about continuing geological change are still accepted today, although modern geologists realize that rates of change have not always been the same.)

Darwin realized the implications of the uniformitarian theory for biology. If the earth itself is dynamic and its present features are the result of a long process of gradual change, Darwin reasoned, could it not be possible that the biological world is also dynamic rather than static?

Fossils of South America

Darwin found his first evidence of change in the biological world in South America. Along the east coast of Argentina, in the mud and silt of river deposits, Darwin discovered a large number of fossil bones literally sticking out of the loose sediment. Among these bones he identified the remains of a giant ground sloth, a huge hippopotamus-like animal, and a species of horse. All three of the fossil forms represented *extinct species*, species not represented by any living individuals. That some species had become extinct suggested to Darwin that not all species were immutable (unchanging).

Even more unsettling to Darwin, however, was the close resemblance of these fossil forms to living species of sloths, hippopotamuses, and horses. Was it possible, Darwin wondered, that these fossil forms might have been ancestral to species found on earth today? If so, the implication was that new species appear on earth as a result of descent from earlier species. Darwin later wrote that his first ideas of evolution began with his observations of these South American fossils.

Galapagos Islands

From Argentina, the *Beagle* slowly progressed southward, rounded Cape Horn, and worked its way northward along the coast of South America to a tiny group of islands called the *Galapagos*. In these islands Darwin discovered a veritable laboratory of evolution. The Galapagos were formed by volcanic eruptions and are relatively isolated, being separated from the South American mainland by some 950 km of ocean. Ancestors of the plants and animals that inhabit the islands came from the South American mainland. They may have been blown to the islands by strong offshore winds or rafted there on floating vegetation. The colonization of the islands by plants and animals was a slow process, but the weathered volcanic rock provided a good substrate for the growth of some plant species, and as seeds germinated and vegetation spread, food and shelter were available for animal species.

When Darwin arrived in the Galapagos in 1835, he found a variety of plants and animals, many of which are found nowhere else in the world. Among the most interesting of the Galapagos creatures were the giant tortoises that gave the islands their name (*galapago* being Spanish for tortoise). Darwin's interest in these animals was aroused when the governor of the islands chanced to remark that he could tell from which island a tortoise had come by observing the shape of its shell. Upon closer inspection, Darwin noticed that not only did the shape of the shell vary from one island to another but that other features of the tortoise varied as well. For example, the necks of the tortoises living in dry areas were longer than the necks of those living in moist areas. Why, Darwin wondered, should so many distinct forms of the tortoise appear in such a limited geographical area?

Darwin later hypothesized that very few tortoises could have arrived at the Galapagos from the mainland, but once there they found little competition for food and no natural predators. Under such conditions, the tortoises became established on the islands and gradually increased in number, spreading out wherever conditions were favorable. Because the physical conditions for survival varied from island to island, the tortoise population of each island developed, over time, individual characteristics that distinguished it from the tortoise populations of neighboring islands. Long-necked tortoises inhabited dry areas where food was scarce, since a longer neck was helpful in reaching high-growing foliage. In moist regions with relatively abundant foliage, short-necked tortoises fared well. Thus, Darwin concluded that the major factor in development of interisland variation among the tortoises was the environmental variation that existed from one island to the next.

Adaptive Radiation

Darwin noted interisland variations among a number of other plants and animals in the Galapagos. For example, on each of the islands there are different species of the genus *Scalesia*, a member of the sunflower, family. Furthermore, in the Galapagos these plants are large, woody, and treelike. In contrast, members of this family that occur elsewhere are herbaceous and considerably smaller.

Darwin also took note of the range of variation among a group of small, drab birds that lived on the Galapagos islands, although he did not collect enough specimens to study them in detail. These relatively

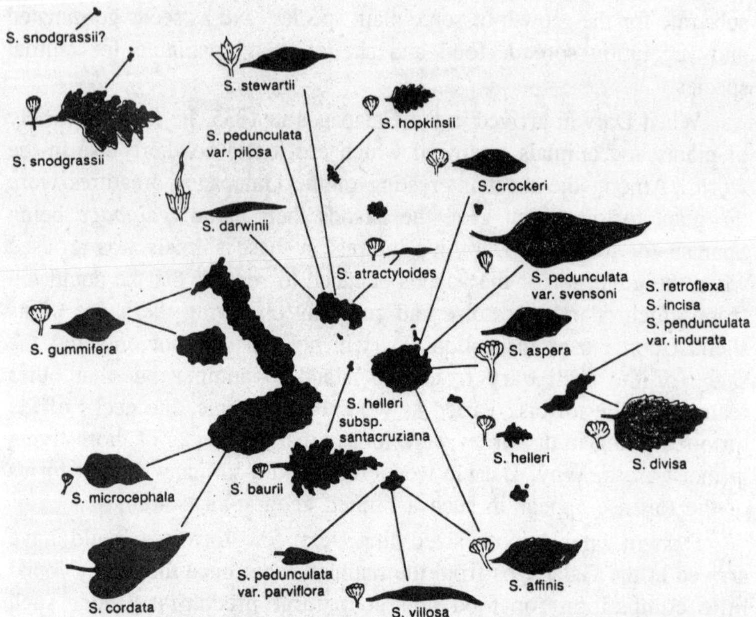

Fig. 3.1. Members of the genus Scalesia on the Galapagos island.

inconspicuous birds have been studied intensively, however, by biologists since that time and are now called "*Darwin's finches*." Like the tortoises, finches were introduced to the Galapagos from the mainland of South America. Also like the tortoises, finches found little competition for food and no natural predators. Under these conditions, the finch population quickly became established.

The original finches were seed eaters, and for a time there was plenty of food and they thrived. But the factors that fostered the initial success of the finches also led to a major problem. Finches are prolific breeders, and with an initial abundance of food and no predators to control population size, the population increased to the point that competition for food and nesting sites became severe. During this period of population increase, the finches spread to all of the Galapagos islands.

On the different islands, the finches underwent evolutionary modification in several directions. Although the finches were primarily seed eaters, they would also eat insects, and those finches that could not obtain sufficient seed to stay alive were compelled to turn to this alternate food source. By doing so, they decreased their competition with the seed eaters and increased their chances for survival and reproductive success. Over time, finches evolved that were insect eaters rather than seed eaters. An important factor in the success of this transition was the lack of competition from other bird species for the insect food supply. In most ecological situations, there are bird species that feed specifically on seeds, on insects, on fruits, and so on. Under such conditions it would be very difficult for members of a seed-eating species to make the transition to insect eating. However, because the Galapagos finches had little competition from other bird species over long periods of time, such evolutionary transitions were possible. Some of the finches even became specialized to other habits, such as cactus eating.

After periods of isolation on separate islands, various groups of finches came to be very different from one another. And these differences involved more than variations in their diet and appearance. These groups of birds, which were descended from common ancestors, had also changed genetically during the period of geographic isolation. They could no longer interbreed when they encountered one another later. They had become reproductively isolated from one another.

Thirteen distinct species of finches occupy the Galapagos islands and one species lives on Cocos Island 700 km to the northeast. Each species is adapted to a different lifestyle. In the Galapagos, finches have filled the ecological roles of warblers, parrots, woodpeckers, and other bird species that would normally occupy these niches in other geographical areas. There are distinct differences among the Galapagos finch species in such traits as plumage, nesting sites, body size, and size and shape of the beak.

The populations of plants and animals of the Galapagos islands illustrate a process modern biologists call *adaptive radiation*. Adaptive radiation occurs when members of a single species enter environments in which there is little initial competition for resources. Under such conditions, after an initial period of adjustment, population size increases rapidly. This leads to intense competition for resources, which in turn may result in fragmentation of the population. After a period of isolation a number of new species may develop; each of these new species occupies a distinct niche, as in the example of Darwin's finches.

Impact of the Galapagos on Darwin's Thinking

Profoundly influenced by his observations in the Galapagos islands and elsewhere on his journey, Darwin concluded that species are not unchangeable and that new species are formed through the gradual modification of existing traits in response to the pressure of both the

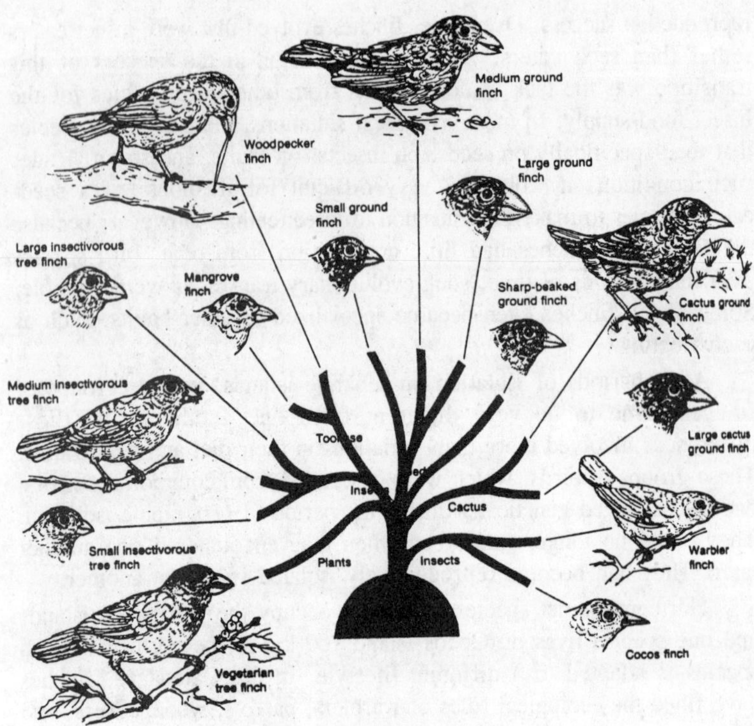

Fig. 3.2. Galapagos finches.

environment and biological competition. Darwin had come to accept the idea of organic evolution—the concept that all species of plants and animals have arisen through descent with modification from previously existing species. However, not every species continues to interact successfully with its environment indefinitely, especially when there are changes in the environment. Change is the rule of nature, and the history of the earth has been filled with many geological and climatological changes. Thus, many species have become extinct. In fact, in the history of life on earth many more species have become extinct than presently inhabit our planet.

Theory of Organic Evolution

After his return to England in 1836, Darwin married Emma Wedgwood and moved to a country house near the village of Downe in Kent. There he spent the next twenty years in research, reading, and developing the idea that living species are the products of series of gradual changes that have occurred through a long period of geological time. This is, in essence, the theory of organic evolution:

that every species of plant and animal is the modified descendant of previously existing species. The concept of organic evolution was not original with Charles Darwin by any means, but he stated it clearly and directly, and systematically gathered evidence in support of the idea.

Darwin realized that this theory required a mechanism that would explain the ultimate source of biological variation and the means by which heritable traits were passed from generation to generation. Although Gregor Mendel was a contemporary of Darwin's, the gene theory of inheritance was not developed until the early years of the twentieth century. Thus, Darwin was unable to explain biological variation. Lacking an understanding of the mechanisms of genetics, Darwin was left with such vague and inadequate concepts as the "blending" theory of inheritance, which attributed the inheritance of traits to a mixing, or blending, of parents' traits. That theory was wholly inadequate to explain the wide range of variation found in any species and could not explain how some traits became more pronounced rather than being diluted generation after generation.

Darwin also realized that he needed an explanation for the mechanism by which new species were formed. The ideas of Jean Baptiste de Lamarck, though later proven wrong, provided Darwin with some food for thought.

Lamarck and Inheritance of Acquired Characteristics

Lamarck developed the concept of inheritance of acquired characteristics as part of his statement of the theory of organic evolution published in his *Philosophie Zoologique* in 1809. Lamarck accepted variation as part of the natural world and believed that the environment played a role in the origin of new species. He thought that new traits were acquired by an organism in response to a need imposed by the environment and that evolution proceeded toward a perfect form.

Lamarck stated that traits could be acquired through the use or disuse of an organ; that is, a body part that is not used by an organism atrophies 'and is lost to future generations, while a body part used extensively is amplified. For example, the Lamarckian explanation for the long neck of the giraffe is that elongation occurred because the ancestors of the giraffe constantly reached into the trees to feed on high-growing vegetation. This repeated stretching of the neck over time resulted in the modern giraffe with its very long neck. A similar argument would be that a son born into a family of weight lifters should have large muscles because his father, grandfather, and great-grandfather had developed bulging biceps by lifting weights.

Lamarck's concept of inheritance of acquired characteristics did not gain wide acceptance in scientific circles, partly because it was *teleological*, that is, it assumed a final goal toward which the process of evolution was directed. Furthermore, although his theory was an attempt to explain how traits are gained and lost, it lacked a basic mechanism of inheritance. And with the eventual rediscovery of Mendel's principles of inheritance, it became apparent that acquired characteristics, which are not genetically based, cannot be inherited. Nevertheless, Lamarck's concept of inheritance of acquired characteristics was one of the ideas about evolution that was considered by Darwin and other scientists of his time.

Artificial Selection

Because he lacked information about genetics, Darwin was not able to provide an adequate explanation for biological variation. He did, however, develop a clear explanation of a mechanism by which evolution of new species can occur.

An activity that has been practiced by the human race for thousands of years provided a clue in Darwin's search for an explanation of evolution. For centuries humans have selectively bred fruits and vegetables, cattle, chickens, and dogs in efforts to obtain greater yields of grain, better milk cows, more productive hens, or dogs more adept at sheepherding or rabbit hunting. Breeders observe the variations that appear in their plants and animals and select for breeding those individuals that best represent the properties they desire in future generations. The success of this *artificial selection* is undeniable, and Darwin saw in artificial selection a model for change in the natural world. What puzzled him was how selection worked in nature. Human beings direct artificial selection, but what force directs selection in nature?

Theory of Natural Selection

As essay on human population written by Thomas Malthus gave Darwin a clue to the solution to this problem. Malthus suggested that human populations have a general tendency to increase in size, but that such factors as fire, war, plague, and famine serve to keep the human population of the earth more or less in check. If it were not for these factors of population control, the human population would increase explosively (a prediction that agricultural, medical, and technological advances of the last century have proven to be accurate).

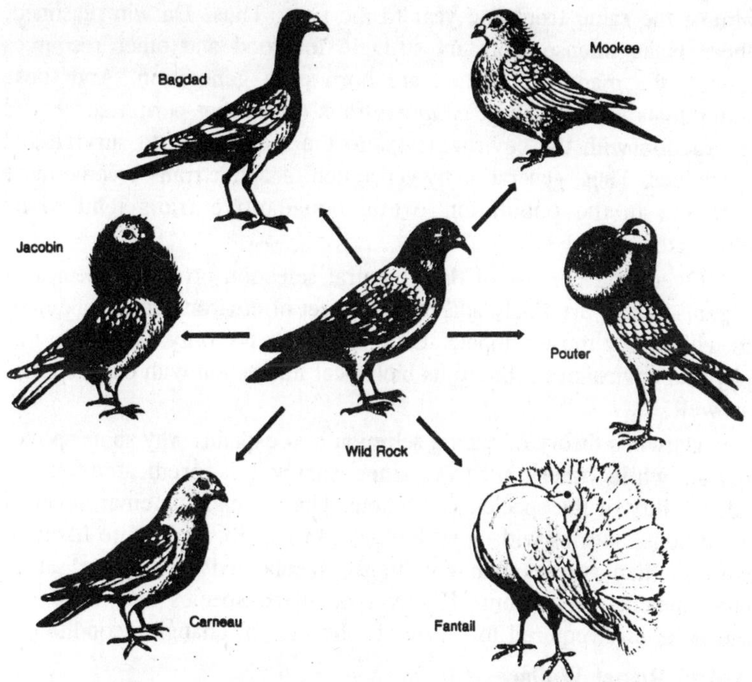

Fig. 3.3. Variation among breeds of the domestic pigeon.

Darwin applied the Malthusian concepts to the natural world and reasoned that every species produced many more young than could survive to reproductive maturity. Of the many new individuals produced by all species in nature each year, most are eaten by predators, succumb to disease, or are unable to obtain an adequate food supply or habitat in which to live. There is a constant struggle for survival. Only some members of any species live to reproduce. Thus, in any generation only the members that succeed in reproducing contribute to the evolutionary future of the species. Darwin recognized that the same factors he had encountered in the Galapagos islands operated throughout the biological world: nature, in the form of environmental variation and biological competition, operates as the selective factor in natural populations.

From these hypotheses Darwin derived his major theory, the *theory of natural selection*. We can summarize the theory of natural selection as follows: All populations of organisms possess the potential to increase at a very rapid rate, but this rate of increase is very seldom observed among natural populations. Instead, the size of populations remains almost the same from one year to the next. Thus, Darwin reasoned, there is an intense, constant struggle for food and other resources among the many young that are born each generation. And those individuals with the most *adaptive* traits (traits that permit successful interaction with the environment) are the most likely to survive and reproduce. Thus, generation by generation, adaptive traits become more frequent in the population, while nonadaptive traits tend to be eliminated.

Over long periods of time, natural selection produces species of organisms that are finely adapted to the set of environmental conditions in which they must compete. Each species is not only adapted to this physical environment, but to its biological interaction with other species as well.

Darwin's theory of natural selection also explains why some species perish while others survive, since survival is predicated on the adaptability of the species as a whole. Under changing environmental conditions, species with great biological variability are more likely to survive than species that are highly specialized to a given set of environmental conditions. Highly specialized species may have lost the flexibility required to survive in the face of changing conditions.

Alfred Russel Wallace

In 1858 Charles Darwin received an essay from the young English biologist Alfred Russel Wallace titled "On the Tendency of Varieties to Depart Indefinitely from the Original Type." Wallace wanted Darwin to review the essay and to decide whether it merited further critical reading by Charles Lyell. Darwin was startled to find in Wallace's short essay the basic elements of his own theory of natural selection. Wallace's ideas were so similar to his own that Darwin felt he should put off publication of his own work and let Wallace have credit for the idea of natural selection.

However, both Lyell and the botanist Sir Joseph Hooker knew of Darwin's long efforts in the development of his theory. They also knew that he had written his own ideas on the origin of species and natural selection in an abstract he had sent to Lyell in 1844 and in a letter to the famed American botanist Asa Gray in 1857. They persuaded Darwin to have these statements of his theory, along with Wallace's paper, presented to a meeting of the Linnean Society in the summer of 1858. Then Darwin hurried on with his preparation of *On the Origin of Species*, which contained his ideas of organic evolution and natural selection, and it was finally published in November 1859.

The similarity of Wallace's independently developed concept of species formation with that of Darwin might seem to be one of the great coincidences in the history of biology, but it is not particularly surprising when you compare the backgrounds of the two scientists. Wallace was a very talented person who had a number of experiences similar to those Darwin had as a young man. Wallace had taken an exploratory journey up the Amazon River and had extensive field experience in the East Indies. He was convinced that living things arose through an evolutionary process and, like Darwin, had pondered the mechanisms of evolution. He also had formulated his ideas about natural selection as a direct result of reading Malthus's essay on population. Despite the similarity of their ideas, Wallace maintained that Darwin should receive primary recognition for the development and formalization of the theories of organic evolution and natural selection. Not only had Darwin's work preceded Wallace's chronologically, but it also was of considerably greater breadth and depth.

However, Darwin and Wallace came to disagree on the action of natural selection in the process of organic evolution. Wallace thought that every trait of an organism was the product of natural selection and therefore must possess an adaptive function. Darwin, on the other hand, maintained that selection was not the only means of species modification and that organisms are complex systems in which certain specific features may have adaptive value while others are nonadaptive. Darwin and Wallace did agree, however, that natural selection is the major force directing the process of organic evolution.

Evolutionary Evidence

New scientific theories often stimulate a great deal of controversy. This has a beneficial effect on scientific progress because scientists with differing views work intensively to find evidence to support their ideas. The theory of organic evolution and Darwin's theory of natural selection initiated such activity, as scientists sought data that would either support or refute these theories. The weight of scientific evidence over the past 100 plus years strongly supports Darwin's basic precepts, and such areas of biological specialization as biogeography, paleontology, comparative anatomy, comparative embryology, genetics, and molecular biology have made especially strong contributions to our modern concepts of evolution.

Biogeography

Biogeography is the study of the geographic distribution of plants and animals. It attempts to explain why species are distributed as they are and what their pattern of distribution reveals about prehistoric habitats and their climates. As Darwin toured the world on the H.M.S Beagle, he discovered that plant and animal species generally are not distributed as broadly as are their potential habitats. Studies in biogeography since the time of Darwin have confirmed this over and over again.

A fundamental conclusion drawn from biogeography is that a new species emerges in one place and then spreads outward from its point of origin. Some species become widely distributed, but they do not spread beyond natural barriers such as those that separate major biogeographic regions. Therefore, although virtually identical environments may exist in different biogeographical regions, they are seldom occupied by the same species. In fact, each major geographic region of the world has its own characteristic sets of plants and animals. For example, in Australia, marsupial (pouched) mammals fill the roles played by placental mammals in other biogeographical regions where marsupials are rare. In addition, the fossil record of each region reflects an evolutionary sequence of biological events distinct from that of all other regions. Within each sequence, the most recent fossil forms closely resemble the living species found in that particular region.

Such observational evidence reinforces the concept that natural selection applied by the environment is the major force that has molded modern species, adapting the life forms of each geographic region into species that are suited for survival within the topographic and climatic conditions of their surroundings. Darwin saw evidence for this when he discovered that species on particular islands are closely related to species on nearby islands, and that island species are generally related to species on the nearest mainland. On the other hand, there is no evidence that supports the existence of a set of "island species" that are characteristically found in island habitats everywhere around the world.

At a more specific level, biogeography provides numerous striking evidences of *convergent evolution*. Organisms in different biogeographical realms, although descended from very different ancestral stocks, possess remarkably similar adaptations to particular habitats. For example, plants of the cactus family (*Cactaceae*) are found in the deserts of southwestern North America and the high deserts of the Andes, but nowhere else. Comparable arid habitats in Africa are occupied by members of the spurge family (*Euphorbiaceae*). This clearly illustrates the molding power of natural selection which has shaped very similar sets of adaptations to similar environments in widely separated parts of the world. The modern concept of continental drift helps to explain the evolutionary histories of the various biogeographic regions and the sets of both fossil and modern organisms found in them.

Paleontology

Information about the history of life on earth is contained in the fossil record, a collection of the remains of extinct life forms. Incomplete as it is, the fossil record does permit scientists to consider events and processes that have taken place through that long history. The study of fossils and the fossil record is called *paleontology*.

Fossil Formation

Of all the organisms that have lived, only a very few have become fossilized, and many of those have later been destroyed by various geological processes. The great majority of dead organisms are devoured by scavengers and decomposed by bacteria and fungi. exposed bones are soon reduced to dust by the combined actions of water, the sun, and wind. The organisms preserved in the fossil record are those few that have been buried by loose sediments very soon after their death. When an organism is buried quickly, oxygen is sealed out and decomposition is greatly retarded. This happens most often underwater and least often in dry upland regions. For this reason, the fossil record contains more aquatic organisms and organisms that lived near water than land-dwelling organisms.

Generally, in order to become fossilized, an organism must possess hard body parts, such as the woody tissue of plants, an external skeleton (shell), or an internal skeleton such as those of vertebrate animals. Soft body tissue seldom survives the rigors of burial in coarse sediments, and organisms without hard body parts are rare in the fossil record.

Fossilization can occur in one of several ways. Water circulating in the sediments surrounding a dead organism may slowly dissolve the water-soluble calcium of shell or bone and leave deposits of another mineral in its place. In this process, called *mineralization*, a durable fossil replica of the original material remains in the sedimentary rocks.

Fossilization also occurs when the original shell or bone is completely dissolved away and removed from the sediment layers by groundwater, leaving behind a perfect *mold* of the organism. At times, a *cast* is formed when minerals are later deposited in the hollow chamber of the mold. Both casts and molds provide good fossil replicas of body surfaces. Other fossils include footprints or skin impressions made in wet mud that later hardens to form a shale.

Interpretation of the Fossil Record

Paleontologists have done a remarkable job of reconstructing the history of life. The first (oldest) life forms to appear in the fossil record were primitive single-celled organisms resembling modern bacteria and cyanobacteria (blue-green algae). In time, more complex cellular forms appeared, followed by multicellular plants and animals. The increase in the complexity of life forms is seen as evidence of an evolutionary sequence—the development of new life patterns continually being put to the test of natural selection. Throughout the long evolutionary process, the great majority of species failed to meet the challenge of adaptation and became extinct.

The fossil record also shows clear evidence of gradual change within single lines of descent. A good example of such change is the evolution of horses. The fossil record of horses covers a period of some 60 million years. The earliest fossil horse was the dog-sized *Hyracotherium* (also known as *Eohippus*), a foot-high animal with four toes on its front feet and three on its hind feet. On the basis of the structure of its teeth, *Hyracotherium* is thought to have fed on forest underbrush. From the variety of horses appearing in the fossil record, we can identify certain trends in horse evolution. First, the molar teeth gradually enlarged and became flatter and higher-crowned, thus making horses better suited for feeding on grasses. Second, the leg bones became fused, providing increased strength for running. Third, the digits of both the forelimb and hindlimb were gradually reduced in number. In the modern horse one digit remains, and the hoof is formed from the nail of this digit. The fusion of the leg bones and reduction of digits were accompanied by an overall elongation of the limbs, which provided greater speed and strength when running over rough ground. In short, the horse became specialized to the life of a grassland grazer. These changes were gradual, and they occurred in a sequence of events, which is depicted in the fossil record.

Horse evolution illustrates that evolutionary changes occur in conjunction with one another, as sets of traits rather than independent traits. The product of this evolution, the modern horse, is a highly specialized animal, molded by natural selection to life on open grasslands. Such specialized animals lack the genetic flexibility to meet changing conditions and would be likely candidates for extinction if their environment were to be modified significantly.

At first glance, horse evolution might seem to represent a "ladder of progress" leading to modern horses. However, if we look again, we see that a number of horse fossils represent evolutionary lines that became extinct. Stephen Jay Gould, one of America's most distinguished evolutionary researchers, suggests that a much better analogy for the fossil record is that of a branching bush. Most "branches" terminated in the past (by extinction), and modern organisms are represented by the tips of a few, still-developing branches.

The fossil record clearly indicates that the major groups of organisms appeared on earth in a sequential pattern. The reasonable assumption is that the older groups gave rise to the younger (more recent) groups, but there are some "missing links" forms intermediate between the major groups of organisms that are absent from the fossil record. However, some clearly intermediate forms have been found. A well-known example of such a transitional form is the fossil species *Archeopteryx*, a pigeon-sized animal that possessed wings with feathers, the beak of a bird, and some skeletal features that partially resemble those of modern birds. However, *Archeopteryx* had reptilian claws, and its skeleton, while somewhat birdlike, was fundamentally reptilian. Specific fossil forms that have been found, such as *Archeopteryx*, may not be the precise species that gave rise to modern forms, but they illustrate evolutionary change.

Another fossil form that had characteristics intermediate between major vertebrate groups is *Seymouria*, an amphibian whose skeletal structure shows features that are a mixture of reptilian and amphibian characteristics. In fact, the skeleton of *Seymouria* is so intermediate in form that it has alternately been classified as a reptile and an amphibian by paleontologists. We will encounter more examples from the fossil record in chapters 37 through 40, in which we survey the diversity and evolution of life on earth.

We should mention at this point that there is some debate as to the tempo of evolutionary change. Many biologists think that the gradual transition in form represented by the fossil record of the horse is typical of evolutionary modification. If this view is correct, then eventually a number of intermediate forms should be found in the fossil record. However, other biologists think that there have been long periods of *stasis*, periods during which species underwent little structural change. According to this view, new species form very rapidly, during periods of environmental stress. Such a relatively rapid production of new forms would leave few transitional forms in the fossil record (*punctuated equilibrium*). This would make it highly unlikely that fossil "missing links" between certain groups would ever be found.

Relative Dating Techniques

To use the fossil record effectively to represent sequential changes that have occurred in the history of life on earth, we must be able to assign ages to fossil organisms. Paleontologists determine the age of fossil-bearing rock layers by two types of techniques: *relative dating* and *absolute dating*.

Relative dating techniques depend on the principle of *superposition*, which was established by Nicolaus Steno in the seventeenth century. This principle states that in any undisturbed sequence of sedimentary rocks, the bottommost layer is the oldest, and the topmost layer is the youngest. This interpretation follows from the manner in which sedimentary layers are formed. When a river reaches the ocean it deposits its sediment load on the ocean floor, and over a period of time hundreds of meters of sediments can accumulate. Eventually, these sediments may be converted into sedimentary rock, with the oldest sediments (and hence the oldest rock) being at the bottom of the rock sequence. Therefore, in any undisturbed sequence of sedimentary rock, fossils found in the bottommost layer are older than fossils found in the layers above. Another method of determining the relative age of fossils was developed in the 1700s by William Smith, a British geologist, surveyor, and engineer. Smith's interest in geology was a practical one, since he was concerned with the physical properties of the rock layers he encountered in his road and canal construction work. Smith wanted to know whether the sandstone he encountered in one place was of the same origin, and thus possessed the same physical properties, as sandstone found in another area. With this goal in mind, he searched for a method to correlate strata from one geographic region with those of another.

Smith noted that in some sequences of sedimentary rock, the kinds of fossils in the bottom layers differed from those in higher layers. Because he had encountered similar fossil groupings in sedimentary layers occurring in widely separated parts of Britain, Smith proposed that all rocks containing similar fossils were of the same geological age. If so, then the fossil assemblage of a particular rock layer would provide the means of correlating the age of rock layers occurring in widely separated geographical regions. Smith's idea proved to be a fruitful one and became known as the principle of *fossil correlation*, a valuable tool used by geologists to this day.

These two principles—superposition and fossil correlation—were the only dating techniques available to geologists before the twentieth century. They were used to painstakingly piece together the fossil record of life in the form of a set of correlations between contents of fossil-bearing rock layers (strata) and their relative ages. No one area on earth contains a continuous sequence of rock strata from the beginning of the earth to the present. Neither does any one location contain a continuous sequence of fossils. The task confronting geologists and paleontologists was to place these scattered rock layers into the proper time sequence. By carefully examining sedimentary layers in many parts of the world and correlating the fossils they contained, a continuous sequence of rock strata and fossil life was attained. The outlines of this record, known as the *geological time scale*.

But there still was no way to assign an absolute age to any single rock layer or fossil. With the discovery of radioactivity and the characteristics of radioactive decay, however, a means was devised to assign specific dates to particular rock layers.

Absolute Dating Techniques

Radioactive atoms have unstable atomic nuclei. When these nuclei break down (decay), they emit characteristic particles or rays. The end result of this radioactive decay is that another kind of atom is formed. For example, *potassium-argon dating* is a method used for determining the age of geological deposits. Radioactive potassium (^{40}K) decays to form ^{40}argon (^{40}Ar). When a sedimentary rock is first formed, it would normally contain ^{40}potassium but no ^{40}argon, since argon is a gas and would be present only if trapped in the sediment during rock formation. With time, however, all of the ^{40}K would decay to ^{40}Ar, but completion of this process would take a very long time. In fact, it would take 1,300 million years for one-half of the ^{40}potassium to change to ^{40}argon. This period of time, called the *half-life*, is the time interval commonly used in *radiometric dating*, dating based on radioactive decay.

Every radioactive substance has its own unique half-life. For example, *uraniumlead dating* is based on the decay of ^{238}uranium (^{23}SU) to ^{206}lead (^{206}Pb), which occurs with a half life of 5,500 million years.

How is this information about radioactive decay and halflives used in radiometric dating? Let us use potassium-argon dating to see how this works.

If a rock sample contains ^{40}K and ^{40}Ar, an exact measurement of the proportion of ^{40}K to ^{40}Ar yields the rock's absolute age. For example, if the rock sample originally contained 1,000 grams of ^{40}potassium, after 1,300 million years it would contain 500 grams of ^{40}K and 500 grams of ^{40}Ar. In another 1,300 million years, half of the remaining ^{40}potassium would have changed to ^{40}argon. Thus, if our rock sample contained 250 grams of ^{40}K and 750 grams of ^{40}Ar, it would have passed through two half-lives and would be 2,600 million years old.

There are several problems with radiometric dating. One is the possibility that in the thousands or millions of years a rock has existed, some of the radioactive isotope or its decay product may have been weathered from the rock. In such a case an erroneous date would be obtained. Another problem is that the best samples of radioactive materials for dating are found in rocks of volcanic origin (igneous rocks). This causes some problems, as fossils are very rarely found in igneous rocks because the heat of the molten material from which these rocks are formed destroys organic matter. Thus, the dating of fossils often depends on finding a volcanic rock layer in close association with a sedimentary rock formation. When this occurs, an absolute date can be assigned to a particular fossil species found in one location, and then fossil correlation can be used to assign an absolute age to other strata containing the same species.

Carbon dating is a means of dating fossils that still contain some carbon material from the living organism. The radioactive isotope ^{14}carbon (^{14}C) has a half-life of 5,760 years, and it breaks down to form a stable isotope of nitrogen (^{14}N). ^{14}Carbon is formed in the earth's atmosphere through the action of cosmic radiation from the sun on ^{14}nitrogen, which is converted to ^{14}C. The ^{14}C in the atmosphere, in the form of $^{14}CO_2$, is incorporated into organic molecules in living tissue through the process of photosynthesis, and when plants are eaten by animals, the ^{14}C is incorporated into the animals' tissues. When an organism dies, carbon exchange with the environment ceases. Any subsequent change in the ratio of ^{14}C to stable carbon isotopes in its tissues is due to radioactive decay of ^{14}C to nitrogen.

This dating technique is based on the assumption that the rate of ^{14}carbon formation in past times was similar to that occurring today and that the ratio of ^{14}C to stable isotopes of carbon (mainly ^{11}C) incorporated into living material was the same in the past as it is now. When the particles (beta particles) that are emitted by the ^{14}carbon in a sample are counted, the total carbon content of the sample can be determined, and the age of the sample calculated. Age is a function of the ratio of ^{14}C to total carbon content; the smaller the ratio of ^{14}C to total carbon content, the older the sample.

One problem associated with the use of the ^{14}C technique is that the relatively short half-life of ^{14}C makes the technique valid only for fossil materials dating back to about 50,000 years ago. Also, it is thought that the influx of cosmic radiation has varied from time to time in earth's history, and thus the rate of ^{14}C formation from nitrogen may not always have been the same.

Using radiometric dating, geologists have estimated that the age of the earth is 4,500 to 5,000 million years, and reasonably accurate dates have been assigned to many of the major events of earth history and to the majority of fossil species appearing in the fossil record. The oldest known fossils, primitive bacterialike organisms, date to approximately 3,000 to 3,500 million years ago.

Comparative Anatomy

Another approach to interpretation in paleontology is *comparative anatomy*. Comparative anatomists try to find similarities and differences among the fundamental structures of living organisms. Their studies have revealed certain basic plans on which groups of organisms are structured. For example, all vertebrate animals share certain basic similarities: a central skeletal axis composed of the skull and vertebral column; a rib cage that protects the heart and lungs, made up of ribs attached to vertebrae; paired appendages; and similarly organized systems of circulation, respiration, digestion, and so on. There are, however, important variations on these basic plans that distinguish the major groups of vertebrate animals from one another.

Comparative anatomists compare many anatomical features of contemporary animals, but the comparative study of skeletal anatomy is most important to paleontologists because fossil evidences of anatomy consist almost entirely of skeletal material.

If all species of organisms have arisen from previously existing species, it would follow that those species that share a recent common ancestor would be more similar genetically. Over time, the number of genetic differences would increase. Thus, as you might expect, distantly related species differ more, both genetically and structurally, than do more closely related species.

Fundamental similarities in structure that have arisen through descent from a common genetic ancestor are called *homologies*. A good example of homology is the vertebrate forelimb. All vertebrates are thought to have evolved from a common ancestor, and despite the fact that bird, whale, and human forelimbs serve different functions, they all possess the same fundamental bone structure.

Analogy is another important concept derived from comparative anatomy. Analogous structures are similar in function and appearance but differ in their fundamental structural plan. The wings of birds and insects are analogous structures. In both cases, the organs are used for flying, and each has a broad, flattened shape that is well adapted to this function. However, the wings of birds and insects are structured differently. The insect wing is a stiffened membrane supported by hard chitinous veins, while the bird wing has an internal bony skeleton covered by skin and feathers. Analogous structures differ in basic form because they have arisen separately. The ability to fly evolved independently in birds and insects. Superficial similarities in the structures of their wings are due to the fact that a broad, flat surface is a prerequisite for flight. Thus, the ability to fly has appeared several times in the history of life. Analogous structures further illustrate convergent evolution, which, as you recall, is evolution from different ancestries toward a common adaptation to a similar life-style.

Comparative anatomy also identifies *vestigial structures*, those structures that appear in a simple, poorly developed form in some organisms, but in a fully developed, functional form in other, closely related animals; or that are poorly developed structures in a modern organism but thought to have been fully functional in its ancestors. Vestigial structures include the rudimentary, functionless eyes of cave-dwelling animals, the wisdom teeth of humans, and the three to five caudal vertebrae present in humans (remnants of the tail structure found in most other mammals).

One explanation for the decline of certain structures is that natural selection favors those individuals in which an unneeded structure is reduced because they expend less energy on its development and maintenance. For example, there is selection against full expression of the genes for eye development in cave-dwelling animals that live in complete darkness. Very slowly, over long periods of time, natural selection may completely remove such genes from populations.

However, it appears that the final instances of expression of certain genes may occur long before the genes are eliminated from the genome. Chickens, for example, have been found to retain genes for the development of tooth enamel, even though the genes are not expressed in modern chickens since they do not have teeth.

Comparative Embryology

Studies of the developmental stages of vertebrate animals reveal similarities that are not apparent in the adult stage. The early embryonic stages of fish, reptiles, birds, and mammals are surprisingly similar. In the past, these similarities were interpreted to mean that among vertebrate animals, "ontogeny recapitulates phylogeny"; that is, the stages of an organism's embryonic development (ontogeny) repeat the evolutionary history of the species (phylogeny). This notion has been discarded in modern evolutionary theory because embryos do not, in fact, pass through stages in which they resemble ancestral adults. Instead, developing embryos of each major vertebrate group pass through some stages in which they resemble the embryos of the other vertebrate groups.

Molecular Evolution

In addition to homologies found in anatomical structures, biochemists have found many homologies at the molecular level. Virtually all living things possess the same genetic material (DNA), use the same genetic code, and conduct energy conversions involving the same molecule (ATP). The fact that practically all living things possess DNA as their genetic material implies that this trait developed very early in the evolution of life and has been passed down through the millenia to the organisms that have followed. It appears that since the DNA method of hereditary transmission and genetic control has proven to be efficient and effective, natural selection has not favored change in fundamental hereditary mechanisms.

Protein homologies have been of special interest to scientists studying evolution. Biochemists can determine the exact amino acid sequence of protein molecules. From this information, the makeup of genes can be inferred because amino acid sequences in proteins reflect nucleotide sequences in DNA molecules. This has made possible genetic studies of evolutionary relationships.

The evolutionary process, as revealed by study of protein molecules, has been very conservative in some cases, but much less so in others. The amino acid sequences of certain histones (proteins closely associated with DNA in chromosomes) have been highly conserved through the course of evolution. One of them has amino acid sequences that are virtually identical in molecules extracted from a wide range of organisms. For example, pea plants and cattle differ by only two of the 102 amino acids in these molecules. Apparently, the proper functioning of histones depends upon the maintenance of a very specific sequence of amino acids; amino acid substitutions in histone molecules that may have appeared as a result of mutation probably were eliminated by natural selection.

Table 3.1. Evolution of cytochrome *c*

Comparison between Different Organisms

Organisms	*Number of differences in amino acid sequences*
Cow and sheep	0
Cow and whale	2
Horse and cow	3
Rabbit and pig	4
Horse and rabbit	5
Whale and kangaroo	6
Rabbit and pigeon	7
Shark and tuna	19
Tuna and fruitfly	21
Tuna and moth	28
Yeast and mold	38
Wheat germ and yeast	40
Moth and yeast	44

Comparison with Human Cytochrome c

Organism	*Number of differences in amino acid sequences*
Chimpanzee	0
Rhesus monkey	1
Rabbit	9
Cow	10
Kangaroo	10
Duck	11
Pigeon	12
Rattlesnake	14
Bullfrog	18
Tuna	20
Fruitfly	24
Moth	26
Wheat germ	37
Mold (*Neurospora*)	40
Baker's yeast	42

However, other proteins have not been so highly conserved, and we can use their amino acid sequences to estimate the closeness of evolutionary relationships among different species. If the amino acid sequences in proteins of two species are very similar, we can assume that these two species are closely related and have relatively recently evolved from a common ancestor. If, on the other hand, the proteins have very different amino acid sequences, the two species are distantly related, and have probably been separated from a common ancestor for a relatively long time.

Data from studies on cytochrome c, an electron carrier molecule in mitochondria, illustate this principle. The amino acid sequences in cytochrome c from humans and chimpanzees are identical, indicating relatively recent divergence of ancestors. Humans and chimpanzees differ from rhesus monkeys by only one amino acid of the 104 that constitute cytochrome c, but they differ from dogs by 11 amino acids, from rattlesnakes by 14, and from tuna by 20.

Using this evidence about the accumulation of protein dissimilarities which have arisen through time by gene mutation, biologists can estimate the relative length of time since ancestors of present-day species diverged from one another.

Molecular Clocks

It is possible, however, that we can learn even more from accumulated changes in DNA nucleotide coding sequences. There is evidence that over long periods of time, gene mutations that alter nucleotide sequences occur at relatively steady rates. This permits biologists to use nucleotide sequence differences as *molecular clocks* that make it possible to calculate how long it has been since the ancestors of two present-day species diverged from a single common ancestor.

By studying the DNA nucleotide sequences of specific genes directly, biologists can analyze these molecular clocks of evolution. An important result of this analysis is the discovery that each gene has it own characteristic rate of nucleotide substitution. This means that, in a sense, there are a number of molecular clocks in an evolving genome, each ticking at its own rate. Nevertheless, phylogenies (interpreted patterns of evolutionary descent) based on molecular clocks generally agree with phylogenies developed by paleontologists using the fossil record and radiometric dating.

There is a cloud on the horizon, however, because of the recent discovery that nucleotide substitutions in certain genes occur at faster rates in some animals (rats and mice) than in others (humans). It is not yet clear what implications this finding has for future applications of the concept of molecular clocks in the study of evolutionary relationshilfs.

ORIGINS: SOME EVOLUTIONARY QUESTIONS

Techniques such as radiometric dating, along with physical evidence from the fossil record and data concerning molecular evolution, provide us with concrete evidence about evolution and the time frame within which organisms have evolved. However, there are several fundamental aspects of the history of life on earth about which we have only circumstantial evidence: the origin of life and the evolution of eukaryotic cells.

Origin of Life

When biologists survey the vast diversity of living things, they often wonder how life originated on earth. There is no way of knowing with certainty, but a number of ideas have been suggested.

The earth probably formed by condensation of dust and gases approximately 4,500 to 5,000 million years ago. The lighter gases, such as hydrogen and helium, were lost as the planet formed because it was too small to hold them by gravity. It is thought that a primitive atmosphere was produced later by volcanic action, but there is considerable disagreement about the exact nature of this primitive .atmosphere. Harold Urey proposed in the early 1950s that the atmosphere was strongly reducing due to the presence of hydrogen. He felt that it was composed of hydrogen, water, ammonia, and methane. More recent studies, however, suggest a mildly reducing atmosphere consisting mostly of water, nitrogen gas, and carbon dioxide, with small amounts of hydrogen and carbon monoxide also present.

As early as the 1920s, the Soviet biochemist A. I. Oparin proposed that organic molecules could be produced from the inorganic constituents of the primitive atmosphere in the presence of an energy source, such as lightning or sunlight. In 1953, Stanley Miller provided support for Oparin's ideas through an ingenious experiment. Miller placed a mixture resembling the probable primitive atmosphere (methane, ammonia, hydrogen, and water) in a closed reaction vessel at 80°C. He then exposed it to electrical spark discharges for a week or more. At the end of this period, Miller discovered that a variety of amino acids and other organic acids had been produced. A number of scientists have conducted similar experiments since the mid-1950s and have shown that a wide variety of biological molecules can be generated using several energy sources. Ultraviolet light or electrical discharges (like lightning, which undoubtedly acted on the earth's primitive atmosphere) can promote these syntheses.

These results are consistent with Oparin's original proposal that early oceans might have contained considerable numbers of organic molecules. Indeed, the early oceans might have resembled a very dilute "soup" or "broth." The formation of this "soup" constitutes the first phase of what has been called *chemical evolution*.

The next phase would have involved the joining of molecular building blocks to yield polymers such as polypeptides, polynucleotides, and polysaccharides. Such molecules are relatively unstable, but they could have been synthesized under gentle conditions through the action of chemical condensing agents like polyphosphates. Polyphosphates are readily formed and are capable of aiding the formation of complex polymers such as polypeptides and polynucleotides. It has also been suggested that reacting molecules became concentrated on the surfaces of clay particles. This concentrating effect would favor polymer formation.

Macromolecules could have subsequently coalesced (come together) to form small droplets containing the polymers in an aqueous matrix. These *coacervate droplets*, as Oparin called them, could have trapped primitive catalysts or enzymes and thus acquired a very rudimentary metabolism. S. W. Fox has heated mixtures of amino acids at 130°C to 180°C to form amino acid polymers, which he called *proteinoids*. When such hot solutions are cooled, these protein oids form small, cell-like structures that Fox has named *microspheres*. Microspheres seem to possess a selectively permeable surface and show cell-like behavior. Fox has speculated that such microspheres might be able to incorporate or develop self-replicating polynucleotides and gradually evolve into protocells.

Some biologists, Francis Crick, for instance, disagree with Oparin and Fox. They think that chemical evolution may have proceeded by the initial development of ribonucleic acid genetic material, rather than by protein coacervates or microspheres. The most primitive nucleic acids might have reproduced themselves and subsequently acquired the ability to direct the synthesis of peptides by using condensing agents like polyphosphates.

Eventually, in one way or another, biological evolution began with the formation of the first true cells. These must certainly have been

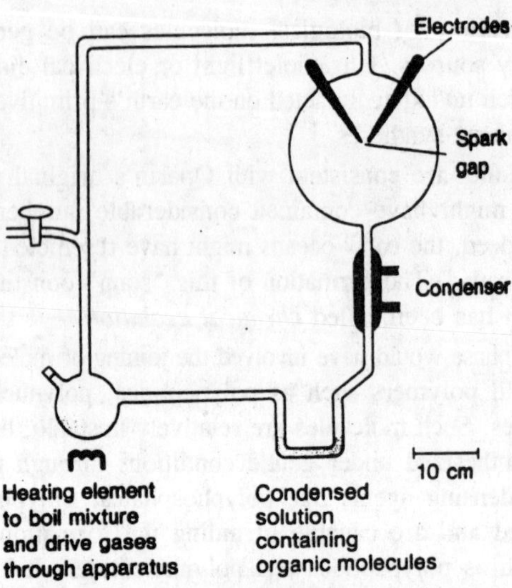

Fig. 3.4. Diagrammatic representation of the spark-discharge apparatus.

fermentative, heterotrophic forms, living at the expense of reduced organic molecules available in the "soup" that surrounded them. Eventually, as nutrients were depleted, the first autotrophs capable of incorporating CO_2 would have arisen. But these first autotrophs were not photosynthetic organisms. They were chemoautotrophs; that is, they derived energy for their synthetic activities not from light, but from chemical compounds available in their environment. These first autotrophs would have had a selective advantage over their heterotrophic predecessors because they would have been able to synthesize organic material from CO_2, rather than having to depend on the dwindling supply of complex nutrient molecules in the environment. This would have been followed by the development of the ability to trap light energy and use it in the process of photosynthesis. With the origin of photosynthetic cells that began to use water as an electron donor and produce oxygen, the primitive environment would have been altered so drastically that the spontaneous development of molecules would have ceased. As oxygen levels increased, cells capable of respiratory metabolism—a much more efficient mode of energy conversion than fermentation—must have developed and flourished as a result of the cells' ability to make effective use of oxygen.

More recently, Berkner and Marshall have attributed the great increase in living forms that took place at the beginning of the Cambrian period, some 600 million years ago, to increases in oxygen concentration.

Between the time at which photosynthetic autotrophs arose (around 3,000 to 3,500 million years ago) and the beginning of the Cambrian period, the oxygen level was too low to screen out ultraviolet radiation through the formation of an ozone layer, which forms from atmospheric oxygen. Early organisms could have lived only deep in the oceans, where they were shielded from the intense radiation striking the earth's surface. When atmospheric oxygen, and the resulting ozone, reached sufficiently protective levels at the beginning of the Cambrian period, life could spread to shallower waters. Finally, about 420 million years ago, the oxygen level rose high enough to provide sufficient ozone to screen out ultraviolet radiation and permit colonization of the land.

Origin of Eukaryotic Cells

The first cells were most likely very simple prokaryotic forms. Eukaryotic cells, however, are much more complex than prokaryotic cells. Eukaryotes contain a nucleus enclosed in a nuclear envelope and an array of membranous organelles such as chloroplasts, mitochondria, and endoplasmic reticulum. The radical dissimilarity between prokaryotic and eukaryotic cells has provoked a great deal of research and speculation among biologists as to how eukaryotic cells might have arisen from prokaryotic ancestors.

Probably the majority of biologists think the *endosymbiotic hypothesis* suggested by Lynn Margulis and others to explain the evolution of eukaryotic cells is correct (at least in broad outline). Symbiosis is an intimate and long-term association between two organisms of different species. Since symbiotic associations, such as lichens, are common today, such associations might have originated and been favored by natural selection early in the development of life. Thus, the mitochondria and chloroplasts of today may have been the endosymbionts (internal symbionts) of yesterday. This concept is the foundation of the endosymbiotic hypothesis.

The fossil record indicates that prokaryotes may have arisen more than 3,500 million years ago, while eukaryotes are thought to have first appeared late in the Precambrian period, about 1,500 million years ago. During the time between the appearance of prokaryotes and the origin of eukaryotic cells, contact among different prokaryotes must have taken place frequently. The first step in the evolution of the eukaryotic cell is thought to have occurred when a large anaerobic amoeboid prokaryote ingested a small aerobic bacterium and stabilized its prey as an endosymbiont rather than digesting it This aerobic bacterium developed into a mitochondrion, the site of aerobic respiration

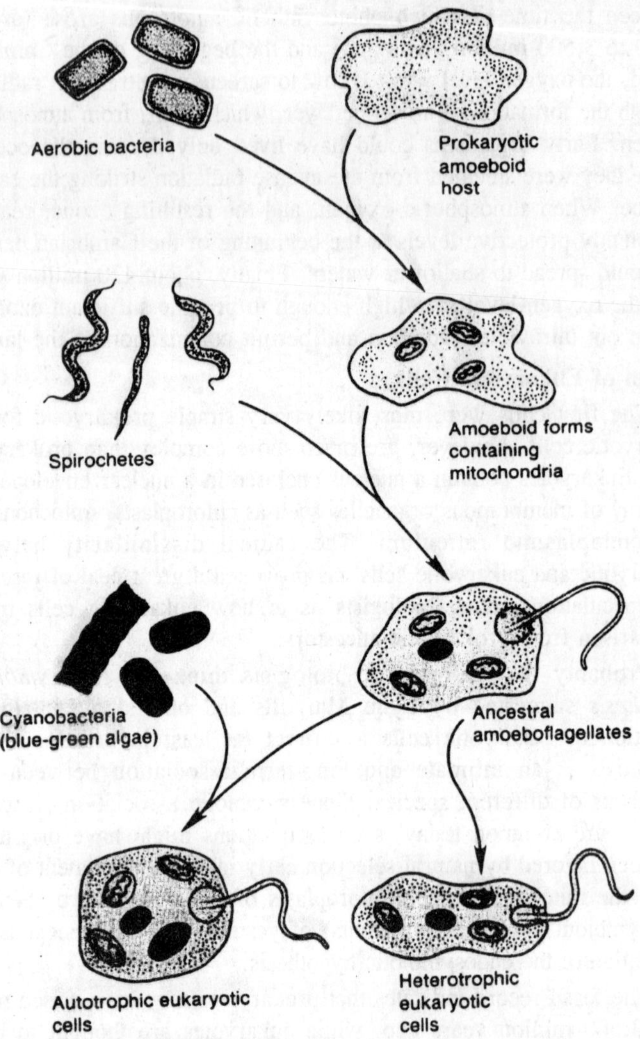

Fig. 3.5. The endosymbiotic hypothesis of how eukaryotic cells first arose.

in eukiryotic cells. Possession of such a mitochondrion-like endosymbiont conferred the advantage of aerobic respiration on its host. Flagella may have arisen through the ingestion of prokaryotes similar to spirochetes. Ingestion of prokaryotes that somewhat resembled present-day cyanobacteria (blue-green algae) could have led to the endosymbiotic development of chloroplasts in a fashion similar to the formation of mitochondria from aerobic, nonphotosynthetic bacteria.

Margulis and others have assembled a considerable amount of indirect evidence in support of their hypothesis:

1. Mitochondria and chloroplasts are similar to bacteria and cyanobacteria in morphology and size.
2. These organelles are self-replicating and contain DNA that more closely resembles the DNA of prokaryotic cells than that of eukaryotic cells.
3. Ribosomes in these organelles are smaller than those in the cytoplasm of eukaryotic cells and are similar in size to the ribosomes of prokaryotic cells.
4. Protein synthesis by bacterial ribosomes, as well as the ribosomes of mitochondria and chloroplasts, is inhibited by chloramphenicol, but not by the drug cycloheximide. In contrast, eukaryotic cytoplasmic ribosomes are inhibited by cycloheximide rather than by chloramphenicol.

The DNA of mitochondria and chloroplasts is much smaller than bacterial DNA, and some organellar proteins can be coded for by nuclear genes. To explain these observations, proponents of the endosymbiotic hypothesis suggest that the endosymbionts must have transferred, over time, some of their genes to the host nucleus and thus relinquished their independence for the sake of the symbiotic relationship.

There are some difficulties with the endosymbiotic hypothesis, and a number of alternative hypotheses for the origin of eukaryotic cells have been suggested. One of these proposes that the prokaryotic cell membrane invaginated (folded inward) to enclose copies of the genetic material, thereby forming several double-walled entities within a single cell. These entities could then have evolved into the eukaryotic nucleus, mitochondrion, and chloroplast.

Regardless of the exact mechanism involved, the emergence of the eukaryotic cell led to a dramatic increase in the complexity and diversity of life on earth. At first, organisms were capable of existing only as independent single cells. Later, however, some evolved into multicellular organisms in which various cells became specialized for many different functions. These multicellular forms became adapted to life in a great variety of environments.

4

MECHANISM OF EVOLUTION

A hospital can be a dangerous place, and the reason has to do with evolution. Antibiotics are used very extensively in modern medicine to fight existing infections and to prevent new ones. This widespread use of antibiotics has acted as a form of selection in an evolutionary sense. Some bacteria possess genes that confer resistance to a particular antibiotic. When the growth of other, nonresistant bacteria is suppressed by the antibiotic, antibiotic-resistant bacteria, though originally few in number, are freed from competition and have more favorable conditions for growth. Thus, the proportion of the bacterial population that is resistant to the antibiotic increases until antibiotic resistance becomes the norm rather than the exception. And all of this is quite likely to happen in hospitals, where antibiotics are used most extensively.

Worse yet, genes for antibiotic resistance commonly are part of plasmids, pieces of genetic material that are separate from the bacterial chromosome. Some bacteria produce multiple copies of plasmids and release them into the environment, where they are taken up by other bacteria. Each cell that takes up a plasmid that includes a gene for antibiotic resistance also becomes antibiotic resistant and is able to pass that resistance on to its descendants. Furthermore, plasmids can be picked up not only by bacteria of the same species that released them, but also by bacteria of other species. Many physicians are trying to limit the use of antibiotics to absolutely vital situations, but the evolutionary race continues. When resistance to a particular antibiotic develops, different antibiotics must be substituted. Often several antibiotics must be used together in the hope that at least one will be effective.

The increase in the proportion of bacteria that are antibiotic resistant is a good example of evolutionary change in the frequency with which genes occur in populations. As we shall soon see, the idea that *evolution involves changes in gene frequencies in populations* is an important aspect of our modern concepts of organic evolution that have developed during the twentieth century.

MENDELIAN GENETICS AND EVOLUTIONARY THEORY

After Mendel's laws were rediscovered in 1900, genetic research was focused on experiments with individual genes and chromosomes using laboratory animals and garden-grown plants. The geneticists who did this research found it difficult to envision the kind of genetic variability that Darwin's principle of natural selection required.

But then, in the mid-twentieth century, a group of biologists, including Theodosius Dobzhansky, G. G. Simpson, Ernst Mayr, and G. L. Stebbins in the United States and Julian Huxley and J. B. S. Haldane in Britain, developed the *synthetic theory of evolution* (also called *neo-Darwinism*). This theory is a synthesis (union) of the theories of organic evolution and natural selection with modern concepts of genetics.

The synthetic theory recognizes that natural populations are much more variable than early experimental geneticists had thought. The primary source of variation is the genetic segregation and recombination that occurs in sexually reproducing organisms. Another important source of variation is gene mutation, which represents the only source of new genetic material in a species.

Under any set of environmental conditions, some phenotypes (sets of observable characteristics) are better adapted than others. The result is differential survival and reproduction of individuals with different phenotypes. Even though natural selection operates on phenotypes, it is genotypes (the genetic constitution of organisms) that determine the phenotypes of individual organisms. Thus, in a very real sense, genes are being selected.

Natural selection can best be explained in terms of its effect on a population's genetic composition. In fact, we can define organic evolution as change in the frequency with which various genes occur in a population. This understanding is the key contribution of the synthetic theory to modern evolutionary concepts. Thus, the focus of modern evolutionary theory is not on individual organisms, but on populations of organisms and gene frequencies within those populations.

Population Genetics

A *population* is an interbreeding group of individuals that occupy a specific geographical area. Because populations may be quite large, numbering in the thousands or millions of organisms, the fate of an individual organism has little influence on the fate of the population as a whole. Thus, modern evolutionary studies concentrate not on the genotypes of individual organisms, but on the *gene pool*, the sum total of all genes available for reproduction in a population. The science of *population genetics* involves mathematical analyses of the events taking place in gene pools, and forms a fundamental part of evolutionary thought.

Hardy-Weinberg Principle

The basis for studying population genetics is a simple mathematical model, a set of concepts called the *Hardy-Weinberg principle* (named for G. H. Hardy, an Englishman, and W. Weinberg, a German, who independently developed and proposed these concepts in 1908). As a mathematical model of the gene pool of a large population, the Hardy-Weinberg principle deals with gene frequencies in that population. It shows how, under defined conditions, the relative frequencies of the alleles of a single gene, considered alone, would tend to remain stable (in equilibrium) from generation to generation. Furthermore, this reasoning applies not just to alleles of a single gene, but can be extended to all alleles of all genes in a population.

Let us use a hypothetical example to illustrate the Hardy-Weinberg principle. To make things simple, we will consider a gene, present in an entire population of sexually reproducing organisms, that has only two alleles—a dominant allele *A* and a recessive allele *a*. In this situation, three genotypes-AA, *Aa,* and as would be possible in the population. Two of those genotypes, *AA* and *Aa,* would produce the same phenotype.

Keep in mind that the Hardy-Weinberg principle deals with population gene pools, not individual genotypes, and is concerned with the frequencies with which alleles occur in the total gene pool.

We will arbitrarily set the frequencies of our hypothetical alleles *A* and *a* at 0.6 and 0.4, respectively. Gene frequencies are always given as decimals, such as 0.6 and 0.4, rather than as the corresponding percentages (60 percent and 40 percent). In every case, the sum of allele frequencies equals 1.0 (100 percent). In our example, 0.6 + 0.4 = 1.0, where allele A makes up 60 percent of the total of the two alleles in the population, while allele a constitutes 40 percent of the total.

If allele A has a frequency of 0.6, it follows that 60 percent of all sperm cells and all egg cells produced in the population carry allele *A*. Similarly, allele *a,* which has a frequency of 0.4, is represented in 40 percent of all gametes produced in the population. These gamete allele frequencies can then be used in a Punnett square to determine the outcome of mating in this hypothetical population.

Because allele A is dominant, 84 percent of the offspring (all those with genotypes *AA* or *Aa*) produced by mating in this population display the phenotype produced by expression of the dominant allele, while only 16 percent display the phenotype produced by expression of the recessive allele.

Will the more frequently occurring allele, allele A (which is also dominant in this case), tend to occur even more frequently in subsequent generations, at the expense of allele a? Do less frequently occurring alleles tend to become lost from gene pools over long periods of time? The answer to these questions provided by the Hardy-Weinberg principle is that if no evolutionary change is occurring in a population, there will be no change in the relative frequencies of the alleles in that population. Even very rare alleles can continue to occur at constant frequencies generation after generation and are by no means doomed to disappear from the gene pool simply because they are rare.

For example, if we analyze the frequencies of the two alleles, A and a, in the gene pool of the generation produced by the hypothetical mating, we see that the relative allele frequencies are the same in the offspring generation as in the parental generation. We could go through this exercise over and over again, producing one hypothetical generation after another, and the relative frequencies of the two alleles *A* and *a* would not change.

The Hardy-Weinberg principle allows us to describe this type of equilibrium mathematically for any population. In such general descriptions, the frequency of one allele is represented by the letter p and the frequency of the other by the letter *q,* and $p + q = 1$. We can insert these symbolic gene frequencies into a Punnett square as gamete frequencies, or we can obtain the same results using simple algebra. You probably recall that the expression $(p + q)^2$ is equal to $p_2 + 2pq + q^2$. Because the sum of genotypes in any gene pool is 1, the following equation can serve as a model of the equilibrium described by the Hardy-Weinberg principle:

$$p^2 + 2pq + q^2 = 1$$

We can illustrate this further by substituting the frequencies of the two alleles (0.6 and 0.4) from the hypothetical gene pool for p and q respectively:

$$p^2 + 2pq + q^2 = 1$$
$$(0.6)(0.6) + 2(0.6)(0.4) + (0.4)(0.4) = 1$$
$$0.36 + 0.48 + 0.16 = 1$$

Thus, the terms of this equation indicate the same frequencies of genotypes as those obtained by the Punnett square method:

$$p^2 = \text{Frequency of AA} = 0.36$$
$$2pq = \text{Frequency of Aa} = 0.48$$
$$q^2 = \text{Frequency of as} = 0.16$$

Calculations become more complex in situations in which more than two alleles are present, but the Hardy-Weinberg principle applies equally well in those situations. In fact, the equations of this model can be applied to all alleles of all genes in the gene pool of a given population.

The take-home message of the Hardy-Weinberg principle is that relative gene frequencies in the gene pool of any population would remain in equilibrium if only completely random genetic processes were acting on it. As gene frequencies do change in evolving populations, however, we can use the Hardy-Weinberg principle as a jumping-off point from which to begin our consideration of the forces that bring about evolutionary change.

Gene Frequencies and Evolution

The Hardy-Weinberg principle presents a mathematical model of genetic equilibrium in a large, randomly reproducing population. The model demonstrates that the genetic recombination that occurs in sexual reproduction as a result of meiosis and fertilization does not by itself produce changes in gene frequencies. But the model of genetic equilibrium demonstrated by the Hardy-Weinberg principle is based on probability and requires that certain conditions be in effect.

These conditions seldom exist in natural populations, however. As a result, the gene pools of natural populations are not found in the stable condition calculated by Hardy and Weinberg. As we said earlier, the genetic composition (gene pool) of natural populations changes from generation to generation, indicating that they are evolving. Evolution involves changes in gene frequencies.

What brings about these evolutionary changes in gene frequency? We can identify four main forces:

1. If a population is relatively small, changes in gene frequency can occur by chance alone. This is called *genetic drift*—change in gene frequency that is not closely tied to the adaptiveness of the traits involved.
2. *Migration* of individuals into or out of the population produces change in the gene pool that is not due to reproductive processes.
3. *Mutation* processes can cause changes in gene frequency.
4. Reproduction is not totally random because not every individual in a population has an equal chance for reproductive success. *Natural selection* may be operating; as a result, certain individuals are more likely than others to contribute their genes to the gene pool of the next generation.

Let us now examine in more detail each of these changeinducing factors.

Table 4.1. Theoretical conditions under which gene pool equilibrium would be maintained according to the Hardy-Weinberg principle

Equilibrium exists if these conditions are met:

1. The population must be large enough so that changes in gene frequency do not occur as a result of chance alone.
2. In-migration or out-migration of individuals must not produce any net change in the gene pool of the population.
3. Mutations must not occur, or there must be mutational equilibrium by which genetic changes in one direction are balanced by an equal number of changes in the opposite direction.
4. Reproduction must be totally random; every individual in the population must have an equal opportunity for reproductive success. Every individual must have an equal chance of mating with any other individual in the population, all matings must produce the same number of offspring, and all offspring must similarly have equal opportunity for reproductive success.

Genetic Drift

The geneticist Sewall Wright developed the idea of genetic drift. Chance events have little effect on gene frequencies in large populations. But in small populations, chance may play a much larger role so that random fluctuations (drift) of gene frequencies occur from generation to generation.

Consider a demonstration of probability in which you place two identically shaped marbles, one of which is white and the other black, in a bag. If you reach into the bag and withdraw a marble, the odds are 50 percent (0.5) that the one you select will be black. If you replace the marble, shake the bag, and draw again, once again the odds are 50 percent that the chosen marble will be black. If you were to repeat this procedure several thousand times, very close to half of your draws would yield the black marble. Yet during your first few draws, you might not pull out the black marble exactly 50 percent of the time. It is only with a large sampling that the black marble will be drawn close to half of the time. It sometimes takes many draws to overcome the inequalities (called *sampling error*) of a limited number of draws.

This example illustrates how chance events can change gene frequencies in a relatively small gene pool; there are not enough reproduction events to "average out" sampling errors. Fluctuations due to chance factors often result in the loss of certain alleles from small gene pools, while other alleles that have been "fixed" in the gene pool become more common. These changes occur independently of natural selection and can result in the loss of certain alleles from a gene pool that could have conferred a selective advantage. As a result, small populations in which genetic drift has occurred tend to have a higher degree of homozygosity than larger populations, which are more variable.

Founder effect

A special case of genetic drift is the *founder effect*. In nature, a few members of a parent population may migrate to a new area and establish a small, interbreeding population (called a *deme*). Because the deme is established by only a few members of the parent population, it is highly unlikely that the gene frequencies of these few individuals would be exactly the same as those of the entire population from which they were drawn. It therefore follows that the gene pool of these individuals' descendants will reflect the gene frequencies of the founder organisms rather than those of the entire original population.

One interesting example of the founder effect in human populations involves the Dunkers, a religious sect of just over 200 people who immigrated to Pennsylvania from western Germany in the early 1700s. Since their arrival in the United States, the Dunkers have maintained strict marriage customs that prohibit marriage outside the sect. They do permit the use of modern medical care and they use modern technology, however. Thus, they probably are subject to the same forces of natural selection as members of the general population.

Under these conditions—that is, a small initial population, genetic isolation from other populations, and no obvious difference in selective pressure-significant differences in gene frequencies between the Dunkers and the United States population as a whole can be ascribed to genetic drift through the founder effect.

In genetic studies, Bentley Glass and his associates at Johns Hopkins University compared the Dunker population with the general populations of the United States and western Germany. Their data show significant differences in the frequencies of the alleles that confer ABO blood types. For example, blood type A occurs more frequently among the Dunkers than in the other two populations, while the 0 type is somewhat more rare among the Dunkers. The B and AB types occur very infrequently. In fact, it was estimated in this study that the IB allele, which must be expressed to produce either type B or AB, had been virtually lost from the Dunker population because most of the few individuals with B or AB blood types were not born in the community but had entered it as converts through marriage to community members.

Certain other physical traits also were examined. Frequencies of some traits, such as middigital hair patterns, hyperextensibility of the thumb, and attached earlobes, were significantly lower among the Dunkers than in the surrounding population. On the other hand, Dunkers have essentially the same incidence of left-handedness as is found in the surrounding population. We can attribute the special set of gene frequencies found among the Dunkers to the founder effect and genetic drift.

Bottleneck effect

Another special case of genetic drift is the *bottleneck effect*, a situation in which a population is drastically reduced by factors other than ordinary natural selection. For example, a particularly harsh winter in the temperate zone will greatly reduce the number of individuals in a population of organisms such as houseflies, that survive to reproduce during the next season. In this way, a large part of the variability in the housefly population may be "squeezed out." Subsequently, relatively few individuals will give rise to all the flies produced during the next summer, and the population may become more homogeneous genetically than it was before.

The bottleneck effect has particular relevance for species that are near extinction. Even if a small number of protected individuals can reproduce successfully enough so that they increase in numbers, new problems may arise. Disease and environmental stress can have devastating effects if the remaining population is very homogeneous. The wider ranges of genetic variability normally found in larger natural populations usually permit at least some individuals to survive crisis situations. An example of such an endangered species is the cheetah. Genetic studies of cheetahs in South Africa reveal an extremely low level of genetic variability. Biologists hypothesize that this may be a result of a population bottleneck in the recent past. Cheetahs may face a dim future even if effective protection and conservation methods are applied.

Migration

Unless there are barriers that prohibit movement of individuals from one population to another, *gene flow* (gene migration) occurs as a result of migration and interbreeding. Gene flow adds to the variability of most natural populations by introducing new alleles. If migration is extensive it can significantly alter gene frequencies.

Mutation and Variability

As you know, variation in natural populations is a fundamental part of natural selection. While mutation is the source of new genes in gene pools, it is sexual reproduction that ensures that different alleles produced by mutation will be combined in many different ways to provide a rich variety of phenotypes in the population. In each generation, sexual reproduction segregates sets of parental genes during meiosis and recombines the genes in new combinations in the offspring.

Mutations can occur as modifications of chromosome structure. Most often, however, mutations are specific changes in individual genes that result from base substitutions in nucleic acid molecules. The frequency with which these changes occur varies because each gene locus has its own characteristic mutation frequency.

Mutation rates of different alleles for the same character are rarely in equilibrium; that is, the rate of *forward mutation*—mutation from the more common allele to the less common allele—is seldom the same as the rate of *reverse mutation*—mutation in the opposite direction. The difference between the two is a *mutation pressure* that tends to produce a very slow change in allele frequencies. For some time biologists thought that mutation pressure was important in determining the direction of evolutionary change, but this no longer seems likely. It would take an enormous amount of time for mutation pressure alone to cause significant changes in allele frequencies. Furthermore, mutations are random events affecting individual genes, but they occur within the context of a whole constellation of genes in a gene pool. Natural selection acts on phenotypes that represent composites produced by the expression of these many genes.

The real importance of mutation is that it is the only mechanism by which new genetic material enters the gene pool, as some mutations produce new alleles.

Modern techniques reveal that mutations involving individual base substitutions in DNA molecules occur so often that it appears that most must be *neutral mutations*, mutations that are without effect. Neutral mutations might, for example, result in the substitution of amino acids in an enzyme molecule somewhere other than in the portion of the protein that makes up the enzyme's active site. Thus, the mutation would not affect the function of the gene's product.

However, of those mutations that are not neutral, a new allele is more likely to prove detrimental than beneficial when the gene is expressed. Since the gene pool of any population is the product of a long period of natural selection during which alleles producing more adaptive traits have increased in frequency and those producing less adaptive traits have decreased in frequency, it is not surprising that completely random mutational events usually produce alleles that are less adaptive than existing ones. However, most mutational events produce alleles that are recessive to the original allele. Such recessive mutant alleles are not expressed as phenotypes until they occur in the homozygous condition. Thus, new recessive mutant alleles are not immediately exposed to the effects of natural selection.

Maintenance of recessive alleles in a gene pool is important because this reservoir of genetic variability may prove advantageous at a later time. Should environmental conditions change, the adaptive value of a specific allele may also change. An example of a detrimental mutant gene that can become beneficial under different conditions is seen in the water flea *Daphnia*. Normally, *Daphnia* thrives at temperatures around 20°C but cannot survive temperatures of 27°C or more. There is, however, a mutant strain of *Daphnia* that requires temperatures between 25°C and 30°C, and cannot live at temperatures around 20°C. Thus, as is often the case, the adaptive value of this mutant condition is dependent on environmental conditions.

Balanced polymorphism

Balanced polymorphism is an evolutionary situation in which several very different phenotypic expressions are maintained in a population without one increasing in frequency at the expense of the others.

Often the selective forces, if any, at work in balanced polymorphism are not evident. Human ABO blood types are one such example. The relative frequencies of the alleles that produce the various blood types are quite stable in populations, implying that the positive and negative selective forces acting on individuals with different phenotypes (blood types) must be in balance, thus producing little selective pressure for change in allele frequencies.

Heterozygote superiority

Selective factors are understood in some cases of balanced polymorphism. One such case involves *heterozygote superiority*, a situation in which natural selection favors individuals that are heterozygous for a gene over either of the homozygous genotypes. Human sickle-cell anemia is a good illustration of heterozygote superiority, but it also shows that there can be differences in the survival value of a specific allele under different environmental conditions.

Sickle-cell anemia is an inherited disease that results from the formation of abnormal hemoglobin molecules in red blood cells. Expression of the S allele produces normal hemoglobin, while expression of the s allele produces abnormal hemoglobin. Individuals with the Ss genotype have approximately 50 percent normal and 50 percent abnormal hemoglobin in their red blood cells.

Individuals with the *ss* genotype suffer from sickle-cell anemia. When they are physically active, their red blood cells collapse and assume a characteristic sickle shape. These sickled cells can block small blood vessels and cut off circulation to a particular region of the body, which in turn can result

in tissue necrosis (death). Individuals with the ss genotype have difficulty with even mild exercise and are generally weak. Many die before reaching reproductive maturity. Heterozygous *(Ss)* individuals sometimes experience problems when they exercise vigorously, but they are generally healthier than ss individuals.

You might expect that the s gene would have a very low frequency in the gene pool. Many ss individuals die young, and heterozygous *(Ss)* individuals also are seriously enough affected at times to be at a selective disadvantage to individuals who are homozygous for the normal hemoglobin allele (SS). However, when geneticists studied populations of black Africans, in which the disease is most common, they found that the recessive allele had a surprisingly high frequency (0.2 to as high as 0.4 in a few areas).

Further research showed that another important factor affects the frequencies of these alleles. In regions where malaria occurs, the heterozygous *(Ss)* genotype had an adaptive advantage over both the *SS* and the *ss* genotypes. Because the malarial parasite, which enters red blood cells, does not inhabit cells that contain the abnormal form of hemoglobin produced by expression of the s gene, individuals with the *SS* genotype suffer a higher death rate from malaria than do individuals of the *Ss* genotype. And since a relatively high proportion of Ss individuals reproduce under these conditions, the frequency of the recessive gene remains greater than would be expected in the absence of the malaria factor. As you might expect, among American black populations, which are not exposed to malaria, the frequency of the s allele is lower than in black populations inhabiting malarial regions.

In the case of the sickle-cell phenomenon in Africa, heterozygotes have a strong selective advantage over either of the two types of homozygotes. Even though both homozygotes are selected against, neither gene is eliminated, since reproduction of the favored heterozygous individuals contributes both genes in equal quantities to subsequent generations. Such heterozygote superiority strongly favors balanced polymorphism.

Natural Selection

Natural selection is a key factor in any changes in gene frequencies within the gene pools of natural populations. Every other factor that causes evolutionary change must be viewed in the context of natural selection. This is because, as Charles Darwin correctly surmised, natural selection is a driving force that causes evolutionary change in living things.

Directional Selection

When the environment changes or when organisms migrate into new environments, natural selection operates to select those alleles that confer traits that are adaptive under the new set of environmental conditions. Because these changes represent a progressive adaptation to a changing environment, this type of selection is called *directional selection*. Directional selection is clearly illustrated by *industrial melanism*, a progressive change in the color of moths that has occurred in industrial areas, particularly in Great Britain and continental Europe. The best-documented example of industrial melanism occurred in the population of peppered moths (*Biston betularia*) in England between the mid 1800s and the early 1900s.

In the mid 1800s, about 99 percent of the peppered moths were light-colored; a darker (melanic) form appeared only rarely. (Genetically, body color in the peppered moth can be treated as a simple dominant-recessive gene relationship, with the allele for dark body color being dominant to the allele for light body color.) Biologists proposed that the greater survival of the lighter-colored moths was due to the fact that it was more difficult for predatory birds to see these moths against light-colored vegetation, while the melanic forms stood out sharply and were quickly attacked and eaten.

The situation was different during the latter half of the nineteenth century, when great quantities of soft coal were burned to fuel the fires of industry. This produced a large quantity of air pollution, and soot settled over the areas surrounding industrial centers, progressively darkening the vegetation. Against the soot-darkened tree trunks, light-colored moths now stood out clearly, while the melanic form enjoyed the advantage of protective coloration. The adaptive advantage of the light-colored form of the peppered moth was lost, and natural selection favored the darker, melanic form.

As a result, by 1900 the peppered moth population of industrialized areas of England was 90 percent melanic and 10 percent light colored. The frequency of the dominant allele had increased over this period of time, while the frequency of the recessive allele had decreased dramatically.

In order to test the hypothesis that the selecting factor in this situation involved predatory birds that feed on peppered moths and that moth body color influences prey selection, H. B. D. Kettlewell of Oxford University reared thousands of peppered moths in the laboratory. He then released equal numbers of the two color phases in two areas: (1) the nonindustrial, unpolluted Dorset area, and (2) the industrial and highly polluted Birmingham area. Each released moth was marked so that when recaptured, Kettlewell could distinguish those he had released from other moths that had inadvertently been captured.

From the unpolluted Dorset area Kettlewell recaptured 14.6 percent of the light-colored moths and 4.7 percent of the melanic form. From the polluted area around Birmingham 27.5 percent of the melanic form were recaptured as compared with 13 percent of the light-colored moths.

Using blinds set up in each area, Kettlewell observed directly the selective predation of peppered moths by birds and even recorded the process on film.

Kettlewell's experiment demonstrated that a moth's body color is indeed a factor in its survival, with the selective advantage of body color varying with the color of the background vegetation. Industrial melanism clearly shows directional selection in which environmental change is accompanied by differential selection of phenotypes, resulting in changes in gene frequencies in the population.

Other Types of Selection

Directional selection generally results in change in one direction in response to changed environmental conditions. However, natural selection can have other effects on gene pools in other situations.

In *disruptive selection*, both categories of phenotypic extremes are favored over the average phenotype in a population. This can happen when a change in conditions removes the selective advantage of the average phenotype. Disruptive selection tends to divide a population into two contrasting subpopulations.

Another type of selection is *stabilizing selection*, which functions when the conditions under which a population is living remain constant over a period of time. Stabilizing selection works against phenotypic variation. It helps to conserve the adaptive fit of the population to its environment by selecting against phenotypes produced by expression of new genetic combinations. Thus, selection is not always an agent of change.

ADAPTATIONS

All biology texts, including this one, are books about adaptations, because all characteristics of living things must be examined in the context of adaptation. An *adaptation* is a characteristic of an organism that increases the organism's *fitness* for life in its environment. Fitness is a measure of the likelihood that an organism will live and succeed in reproducing. Fitness, therefore, determines the odds that an organism will make a genetic contribution to the next generation and is a critical measure of evolutionary success.

Each living thing possesses a battery of adaptations that contribute to its fitness. We can illustrate the complexity and diversity of adaptation by examining some examples of special types of adaptations, such as those involved in certain interactions between organisms of different species.

Cryptic Coloration and Mimicry

For most organisms, one of the most urgent and continuing threats to survival and eventual reproductive success is the danger of attack by predators. Many adaptations associated with predator avoidance are truly remarkable. For example, some organisms are camouflaged by *cryptic coloration* (“*hidden coloration*”) that makes them virtually undetectable when they are in position against their normal background. Other animals, like the well-known chameleon, undergo complex color changes that improve their “match” with the background as they move from one location to another.

Through *mimicry*, some organisms, instead of being hidden from the eyes of potential predators, present a showy but misleading appearance resembling that of another organism that predators normally avoid. The similarity between the mimic and the model may be so close that predators avoid both to nearly the same extent.

Coevolution

Coevolution is a situation in which the mutual evolutionary interaction between two species is so intense that each exerts a strong selective influence on the other. We have already described some examples of adaptations resulting from coevolution. For example, rumen microorganisms and their grazing hosts are products of a long coevolutionary process.

Some particularly striking products of coevolution are the relationships between certain flowering plants and the animals upon which they depend for pollen transport from flower to flower.

Most flowers pollinated by bees are showy and bright, and all of them are open during daytime hours, when worker bees do their foraging. Bee-pollinated flowers tend to be blue or yellow; few of them are red, since bees are blind to red colors. Furthermore, most of them have sweet, aromatic odors because bees depend on smell to locate nectar-containing flowers. Bee-pollinated flowers also have petal arrangements that provide bees a place to land.

Hummingbirds, on the other hand, see red well but blue only poorly. Because hummingbirds have a poor sense of smell, aroma is not an important factor in attracting them to flowers. Like bees, hummingbirds forage in the daytime, but they require no landing perch. A hummingbird hovers in front of a flower and inserts its long beak and tongue, which is specialized for sucking, into the nectar-containing part of the flower. Characteristically, humming-bird-pollinated flowers are red or yellow and many of them are odorless.

Moths, such as sphinx moths, also hover as they feed on nectar from flowers, but they feed during evening or nighttime hours. Thus, the flowers that moths pollinate are ones that remain open at night, and in most cases are light colored.

The scarlet gilia (*Ipomopsis aggregata*), a plant that grows on mountainsides in the southwestern United States, changes flower color during its growing season and thus attracts the particular pollinators that are present at different times during its flowering period. Early in the season, when hummingbirds are present, most high-altitude gilias produce bright red flowers. Later in the season, after the hummingbirds have left the area, many of the plants produce lighter-colored flowers that attract moths, which continue to feed for several hours after dark. Ken Paige and Thomas Whitham, who studied this adaptation, reported that in late season, the percentage of gilia plants with light-colored flowers that set fruit (indicating successful pollination and fertilization) was significantly higher than that of redflowered gilias.

Even behavioral adaptations of pollinating animals show evidence of the coevolutionary process. Bees, for example, tend to feed "faithfully" on one kind of flower at a time. This is adaptive for bees because plants of many species flower practically in unison, and by tending to seek more flowers like those in which they have found nectar, the bees' search for nectar is more efficient. If bees tend to move from plant to plant of the same species, pollination is effectively accomplished. A hovering moth likewise moves systematically from flower to flower in a flower bed. It pauses before a blossom long enough to unroll its long, sucking proboscis and extend it deep into the flower to feed. Then it rerolls its proboscis and moves on, repeating the process again and again until every flower has been visited.

Symbiosis

Some interactions between organisms that are products of coevolution are much more intensive and permanent than the temporary encounters between flowers and animal pollinators. *Symbiosis* (literally "living together") is a general category of relationships in which one organism lives in an intimate and continuing association with another. Symbiotic relationships range from the exploitative relationship of parasitism in which one member benefits at the other's expense to mutualistic relationships in which both organisms benefit from the association, as in the case of the algae and fungi that make up lichens. In these and the many other symbiotic relationships in the living world, the impact of coevolution is apparent.

Species and Speciation

The changes in gene frequency we have considered so far, such as the increasing incidence of antibiotic resistance in bacteria, are changes that occur within the gene pools of populations and constitute what is called *microevolution*. On occasion, however, sufficient genetic change has accumulated so that a new species has been formed. This is called *macroevolution*. Let us now turn our attention to the species concept and the mechanisms by which macroevolutionary change (the production of new species) takes place.

Species Defined

A *species* is a population of organisms that may display a range of phenotypic (and genotypic) variation, but that nonetheless represents a biological entity distinct from all other populations. One definition of species states rather simply that all members of a species are more like one another than they are like members of any other species.

However, the concept of *reproductive isolation* gives us a more satisfactory definition of a species. This concept defines a species as an interbreeding population that possesses a genetic identity because its members do not interbreed with members of any other species. Reproductive isolation may thus serve as a working criterion for species definition.

Reproductive isolation

Certain factors prevent successful interbreeding between individuals of different species, and thus maintain reproductive (and genetic) isolation of the species. There are a variety of these *reproductive isolating mechanisms*.

In animals, certain premating isolating mechanisms operate to prevent mating between members of isolated populations. One such mechanism is *mechanical isolation*. For example, there may be differences in the reproductive structures of members of two populations that make copulation difficult, if not impossible. *Habitat isolation* occurs when two species inhabit the same general area, but occupy very different specific habitats. Thus, members of these two species seldom venture into each other's habitats. *Seasonal isolation* is a third premating mechanism. In this case, different species may become sexually active at different times of the year. A final premating isolating mechanism is *behavioral isolation*. The various songs, calls, display patterns, and courtship "dances" of birds and other animal species attract mates of that species but eliminate members of other species as possible mates, thus preventing gene flow between them.

Prefertilization reproductive isolation mechanisms function in plants also. For instance, when pollen grains of one species fall on the stigmas of another species, pollen tubes usually do not develop. Cross-pollinations among such species cannot result in fertilization.

Post-mating isolating mechanisms operate after copulation has taken place. *Gamete mortality* occurs in matings between members of different animal species with internal fertilization. Sperm produced by males of one species may not be able to survive in the female reproductive tract of the other species, or the zygote may not be able to begin normal development. *Hybrid inviability* refers to the death, anytime before reproductive maturity, of hybrid offspring resulting from matings between members of different species. Even if hybrid offspring survive, there may be *hybrid sterility*, in which hybrid offspring are healthy and survive to reproductive age, but are incapable of producing offspring when mated with either parent type or with another hybrid. A familiar example of hybrid sterility is the mule, the hybrid offspring of a horse and a donkey. Mules are strong animals, well suited to certain work tasks, but despite scattered reports of their producing offspring, mules generally are sterile.

In situations characterized by hybrid inviability and hybrid sterility, members of two populations may mate successfully and produce offspring, but in neither case is a new genetic line of descent established. These isolating mechanisms function either because of significant genetic or chromosomal differences between the parents. We will note later, however, that barriers to hybridization between species are much less rigid in plants than in animals.

There are, however, some short-comings in the reproductive isolation criterion of species definition. For example, because it is impossible to test for crossbreeding among all members of any two natural populations, it is impossible to state with absolute certainty that complete reproductive isolation exists.

The reproductive isolation criterion also gives us no help in studying species known only from the fossil record. Structural studies remain the major criteria for fossil species definition.

Moreover, the reproductive isolation criterion cannot be used to define organisms that reproduce asexually.

Yet another problem arises from studies of widely distributed species such as the leopard frog, *Rana pipiens*, of North America. John Moore and others have experimented with crosses between individuals from different geographic regions. They have found that frogs from adjacent regions anywhere in the range of the species can successfully interbreed with one another. However, when crosses are made between individuals from distant points, such as Florida and Vermont, the embryos produced by these matings develop abnormally and invariably die. This illustrates an important aspect of the "species problem." There is a potentially continuous chain of successfully interbreeding populations of leopard frogs between Florida and Vermont, and yet, the distant populations at the "ends" of the chain are not able to interbreed successfully.

It is probably apparent by now that the criteria currently used for species definition are not altogether clear-cut, and it is not surprising that biologists are not in total agreement as to how the criteria should be applied.

Speciation

What are some of the means by which populations become reproductively isolated and thus established as separate new species?

Allopatric speciation

A very important mechanism involved in species formation is *allopatric* ("other countries") *speciation* (also called *geographic speciation*). This occurs when a natural barrier, such as a desert, river, chain of mountains, or ocean, forms and physically separates a species into two or more populations. Because these populations can no longer contact each other, they are prevented from interbreeding, and there is no gene flow between them.

The way in which geographic separation of populations can lead to speciation. In stage 1, species A is continuously distributed over a region that, except for a small desert, is uniformly cool and moist. At this stage there is a large population in which mating is random and there is no interruption of gene flow. Then the climate of the region begins to change—a dry wind starts to blow across the desert, causing an extension of very dry conditions and a considerable warming of another area. Under these conditions, the homogeneity of species A is disrupted so that in stage 2 there are demessmaller, restricted interbreeding subpopulations—that are subjected to different environmental conditions. The gene pools of these demes may differ slightly, but they are not so divergent that gene flow between them could not occur. At stage 2, the environment of the A deme continues to be much as it was; A_1 is subjected to dry, semiarid conditions; and A_2 is subjected to a warmer environment.

By stage 3 the desert has extended, and the A_2 deme has disappeared because its habitat has disappeared. The desert has become a physical barrier that completely separates A from A_2, removing the possibility of gene flow between them. These two demes are now allopatric; they are separated by a physical barrier, the desert. From this point onward, A and A_2 are separate populations that experience different selective forces. It is quite possible that genetic drift may also contribute to differences between the two demes, especially if the original number of individuals in one of them was small. In time, sufficient variation may accumulate in the two gene pools for reproductive isolation to occur, so that even if these populations were to be rejoined, they would not be able to interbreed. Thus, new species have evolved.

More commonly, allopatric speciation probably occurs without major geographical alterations in the environment of a given species. Ernst Mayr points out that demes can become isolated at the periphery of the range of a species in locations and situations where they are unlikely to interbreed with the larger central population of the species. Genetic drift and local selective forces may lead to new species formation in such populations, just as in those separated when absolute geographic barriers develop, such as in our example of the spreading desert. Mayr has named this kind of speciation *peripatric* ("at the edge of the country") speciation.

Sympatric Speciation

New species also form through *sympatric* ("same country") speciation; that is, new and different species originate from populations that occupy the same geographic area. In this case, no geographic barriers are involved.

In plants, sympatric speciation can occur by the development of *polyploidy*, the possession of more sets of chromosomes than normal. This is caused by abnormal cell division. If chromosome replication is not followed by separation of the duplicated chromosomes, a polyploid cell with multiple sets of chromosomes is produced. When this happens during meiosis, it can eventually result in the development of new plants that have extra sets of chromosomes. Their different chromosome number makes them sexually incompatible with, and thus reproductively isolated from, the parent species. But they can interbreed among themselves; they are a new species.

Such accidental chromosome duplication also makes possible the formation of new species by hybridization between existing plant species.

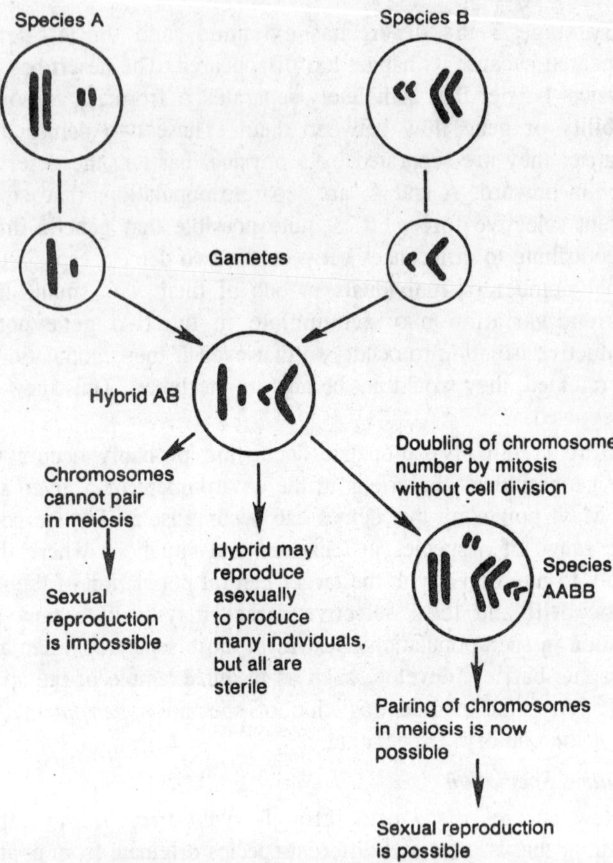

Fig. 4.1. A hypothetical example of hybridization between plant species illustrating the role of chromosome duplication in converting a sterile hybrid into a new species capable of sexual reproduction.

As we noted earlier, in most cases hybrid offspring of interspecific matings are sterile. One reason for this is that the cells of a hybrid may not have paired sets of homologous chromosomes, but rather single chromosomes received from its two parents that do not form homologous pairs. When the hybrid matures, normal meiosis is impossible and sexual reproduction is prevented. However, this barrier is removed if mitosis without cell division occurs during the growth of a hybrid plant so that its cells have paired sets of chromosomes. Then, at maturity, normal meiosis can occur and the organism is fertile and can interbreed with others like itself, but not with members of either of its parent species.

Such chromosomal events occur relatively frequently in plants. In fact, about half of all flowering plant species in existence today arose by polyploidy. Many of these, including economically important plants such as wheat, originated as hybrid polyploids.

There is some debate, however, as to whether or not truly sympatric speciation occurs in animals. No habitat is entirely homogeneous, and at times, small subpopulations of an animal species may become highly specialized to localized environmental conditions. Can enough differences accumulate between these subpopulations to produce reproductive isolation, even though members of these subpopulations have the opportunity to interbreed? There is some evidence in certain insect species that specialization of subpopulations to utilize very different food sources might indicate that these subpopulations are on their way to becoming separate species. However, many questions about speciation within truly sympatric animal populations still remain.

Tempo of Evolution

Early evolutionists often thought of evolution in terms of gradual change from one species to another. There is evidence of such straight-line descent, called *anagenesis*, in the fossil record. However, a more common pattern of speciation appears to have been *cladogenesis*, formation of a new species by branching from a previously existing species. In recent years an interesting debate has developed over the tempo of this evolutionary branching. The traditional view has been that species are formed as a result of a slow, gradual accumulation of adaptive traits in populations through a selective process that requires many thousands or even millions of years for the separation of new species. This is the concept of *phyletic gradualism*. Recently, however, a number of biologists, notably Stephen Jay Gould and Niles Eldredge, have advanced the hypothesis of *punctuated equilibrium*, which suggests that evolution has involved a series of relatively rapid speciation events. They argue that throughout the greater part of its existence, a species displays very little change. This equilibrium is suddenly interrupted by rapid evolution of a small, local population at the periphery of the range of a species, and the process of branching into a separate new species occurs within a relatively short period of time.

Proponents of punctuated equilibrium suggest that the fossil record supports their interpretation of the evolutionary process because it shows long periods during which species remained essentially unchanged. They also point out the lack of transitional forms between taxonomic groups.

Scientists who support phyletic gradualism have held that transitional forms are missing because relatively few organisms have been preserved in the fossil record, and it is not likely that, by chance, a large number of intermediate forms would be found. However, the supporters of punctuated equilibrium suggest that transitional forms are not found because there simply may not have been transitional forms. In other words, species formation occurs so rapidly that the long sequence of intermediate types predicted by phyletic gradualism were not a part of species formation.

One way that evolution could proceed by such leaps would be by the occurrence of a mutation, such as a major chromosomal reorganization that drastically alters early development so that the organism produced is extremely different from its parents. Under normal conditions, most such unusual individuals would probably die relatively young because they would not be well adapted to their environment. But occasionally, one or more would be very much better suited than ordinary members of the population for a rapidly changing or highly stressful environment.

Rapid speciation also might take place by other means, such as the occurrence of periods of very strenuous selection that favor rapid evolutionary change. Regardless of the mechanism by which it might occur, the concept of punctuated equilibrium has sparked a great deal of debate among biologists.

Unfortunately, some observers outside the field of biology misinterpret debates such as the one concerning phyletic gradualism and punctuated equilibrium. They mistakenly conclude that biologists "are beginning to have doubts about evolution." These biologists are *not* expressing doubts about the overwhelming evidence that indicates the occurrence of an evolutionary process. Rather, they are debating the way in which that evolutionary process proceeds and the mechanisms by which new species arise. Such debates actually stimulate biologists to reexamine the existing data and to seek new information regarding the process of evolution, thereby encouraging scientific progress.

Modern biologists explain the process of organic evolution in terms of changes in gene frequencies occurring in gene pools of populations. The Hardy-Weinberg principle provides a mathematical model of genetic equilibrium against which evolutionary changes in gene frequency can be measured. Gene frequency changes may be caused by genetic drift, migration, and mutation. Natural selection is a major force causing gene frequency change.

Each organism possesses a set of adaptations to its environment. Some adaptations are particularly striking because of their involvement in interactions between organisms of different species. Cryptic coloration and mimicry are adaptations that function in avoiding predation. In coevolution there is an intense mutual evolutionary interaction between two species in which each exerts a selective influence on the other.

A species is a population of organisms that is reproductively isolated from members of other populations. Species may also be defined on the basis of structural similarities and differences.

Allopatric speciation is a very important mechanism for species formation. A population is divided into smaller units and changes in gene frequencies accumulate within their individual gene pools until reproductive isolation is established, at which point new species have been formed.

It is also possible for species to form by sympatric speciation, as when the development of polyploidy reproductively isolates plants with different numbers of chromosomes.

There is disagreement among biologists concerning the tempo of species formation. Does speciation occur as a result of long-term, gradual accumulation of adaptive traits? Or are there relatively short periods of rapid species formation that punctuate longer periods of comparative equilibrium? Seeking answers to these questions helps to clarify the nature of the evolutionary process.

5

EVOLUTIONARY RATES

A genetic system is defined by its basic mode of inheritance (bisexual versus unisexual, diploid versus haploid) and, when referring to a specific phenotype, by its genetic architecture (number and relative importance of loci, the pattern of dominance, epistasis and pleiotropy, linkage relationships, etc.). Both influence the rate of evolution, but this chapter will focus only on basic mode of inheritance.

The interaction of cladogenesis and mode of inheritance is illustrated by examining the impact of founder events in *Drosophila*. Basic theory predicts that X chromosomes (a haplo-diploid genetic system) should be more sensitive to founder effects than autosomes (a bisexual diploid system). An examination of the genetic basis of evolved differences in both natural and experimental founder populations reveals a strong preferential involvement of X-linked loci over autosomal loci. Further evidence for the increased sensitivity of X chromosomes to founder events is provided by the fact that X chromosomal inversions are preferentially fixed over autosomal inversions in Hawaiian *Drosophila*, a group for which founder events are important, but there is no preferential fixation in continental *Drosophila*. Basic theory also predicts that mitochondrial DNA (mtDNA) should be more sensitive to founder events than nuclear DNA. This prediction is tested by contrasting two lineages of Hawaiian *Drosophila*, one for which founder events have apparently been important in speciation and the second for which founder events were not important. The pattern of nuclear versus mtDNA evolution is discordant between these two lineages, and the difference is shown to be due to a greatly accelerated rate of mtDNA evolution in the lineage subjected to inter-island founder events.

The importance of anagenesis in determining rates of evolution is illustrated by contrasting rates of evolution in mtDNA and nuclear DNA in the hominoid primates. The human lineage has had a large increase in generation time and a simultaneous expansion of population size over at least the last two million years. Theory predicts that these trends should cause an evolutionary slowdown in all genetic systems. This prediction is supported by a statistical analysis that indicates that mtDNA, nuclear DNA and karyotypes all have slower rates of evolution in humans than in African apes.

In summary, the rate of evolution of genetic systems is sensitive to both cladogenesis and anagenesis, but the degree of sensitivity may differ among the genetic systems. The heterogeneity in evolutionary rates does not mean that molecular data cannot be used to estimate divergence times; rather, it is shown that more accurate estimates are possible when rate heterogeneity is taken into account. Moreover, patterns of differences and uniformity among different genetic systems can therefore be used to make inferences concerning the events that have influenced the evolution of the group under study.

A genetic system is defined by its basic mode of inheritance (bisexual versus unisexual, diploid versus haploid) and, when referring to a specific phenotype, by its genetic architecture (number and relative importance of loci, the pattern of dominance, epistasis and pleiotropy, linkage relationships, etc.). Templeton (1982a,b) has shown that the type of genetic architecture strongly interacts with the mode of speciation and the type of anagenetic evolution in determining rates of evolution. Because genetic architecture has been discussed elsewhere, this chapter will focus only on basic mode of inheritance and its relationship to evolutionary rates.

Previous work on the impact of basic mode of inheritance on evolutionary rates has traditionally dealt with the problem of the evolution of sex. Thus, there is an extensive literature comparing rates of evolution in sexually reproducing, diploid species versus asexual haploids or parthenogenetic diploids. However, intraspecific comparisons are also possible because most organisms have several different genetic systems rather than just one. For example, in *Drosophila* and mammals, four distinct genetic systems are readily identified. First, there is the autosomal system that is inherited as a bisexual diploid. Secondly, the X chromosome is bisexually inherited, but is diploid in females and haploid in males. Thirdly, the Y chromosome is a paternally inherited haploid and finally, the mitochondrial DNA (mtDNA) is inherited as a maternal haploid. Unlike comparisons across species, the genetic systems imbedded within the same species obviously share many common evolutionary constraints; yet, their differing modes of inheritance allow these constraints to display a variety of potential impacts. The purpose of this chapter is to illustrate the impact of both cladogenetic and anagenetic evolutionary constraints on the rates of evolution of diverse genetic systems, showing that this impact across genetic systems imbedded within the same species can be homogeneous in some cases but heterogeneous in others.

Cladogenesis and Evolutionary Rates

Founder Events and Genetic Systems

The impact of cladogenesis on the evolutionary rates of diverse genetic systems will be illustrated by speciation events induced by founder effects. Carson and Templeton (1984) have recently reviewed the evidence and theory for founder-induced speciation, and they have concluded that speciation events are likely only when the founder event is followed by a rapid and large increase in population size.

Nei et al. (1975) examined the genetic impact of such founder-flush events. They showed that founder events of size 10 have very little impact on overall levels of genetic variation and fixation when followed by a rapid increase in population size. Even the most extreme founder event possible, a population of two individuals, carried over a minimum of 65% of its genetic variation. The role of the founding number was investigated in more detail by Senner (1980). Restricting attention only to the case when the founding event was followed by a large increase in population size, Senner found an inflection point at about four individuals. Above four, most genetic variation is preserved and little fixation occurs. Below four, the amount of fixation is greatly influenced by the actual number of founding individuals, with decreasing numbers of founders resulting in increased fixation.

It is important to realize that both the Nei et al. (1975) and Senner (1980) models refer to bisexual, diploid genetic systems. Hence, the inflection point of about four individuals actually translates into an inflection point of about eight haploid genomes. A founder event consisting of N individuals will consist of diverse numbers of haploid genomes for the various genetic systems imbedded within those individuals. For example, much of the speciation in the Hawaiian *Drosophila* is most likely due to founder events involving a single gravid female. Suppose the female had been inseminated by M males, so that the founder size in terms of number of individuals is $N = M$

+ 1. Then, the number of autosomal genomes present in the founder population is $2M + 2$, the number of X chromosomes is $M+2$, the number of Y chromosomes is M, and the number of mitochondrial genomes is 1. Obviously, the various genetic systems can show extremely different sensitivities to the same founding event. For example, if $M = 3$, the autosomal system should be relatively insensitive to the founding event, but the X, Y and rntDNA will show progressively increasing sensitive in terms of loss of genetic variation and increased fixation rates.

X Chromosomal versus Autosomal Genetic Systems

Experimental founder events

For several years, I have been investigating the capacity for parthenogenetic reproduction in a natural, and normally bisexually reproducing, population of *Drosophila mercatorum*. As argued elsewhere, these screenings for parthenogenetic capacity constitute an experimental investigation of founder events that result in the most extreme inbreeding possible. The parthenogenetic lines were established from the virgin female offspring of single wild-caught females. Hence, the sources of genetic variation available for selection in the establishment of any given parthenogenetic strain is the genetic variation brought into the laboratory by a single gravid female. Moreover, most wild-caught females are multiply inseminated, with the probability of two offspring of the same female having different fathers being 0.32. Hence, these parthenogenetic strains are established from isofemale founding lines for which the founder effect will be much greater for the X chromosome than for any autosome.

Once a parthenogenetic strain is established, a sexual analogue can be bred. Hence, one can focus directly on the genetic impact of these experimental and extreme founder events rather than parthenogenesis *per se*. The most extensive work to date has been done on one of the first parthenogenetic lines established, K28-0-Im. Several isolating barriers evolved in this line, so it is an ideal experimental model of speciation. First, this line is characterized by extremely strong premating isolation from certain other strains of *D. mercatorum*. Chromosomal contrasts have been performed that demonstrate that the sexual isolation depends upon a strong epistatic interaction between two chromosomes: the X chromosome and the acrocentric II autosome. No other chromosomes are involved.

Given that mating occurs, other isolating barriers are unmasked. Many of the F_2 and backcross males lack motile sperm and are sterile.

The genetic basis of sterility is similar to that of premating isolation; sterile males are produced by simultaneous hemizygosity of the K28-0-Im X chromosome and homozygosity for the K28-0-Im acrocentric II autosome. Although the same chromosome pair is involved in both premating isolation and male sterility, these traits are separable genetically.

Finally, there is an F_2 breakdown in viability associated with the phenotypic syndrome known as *abnormal abdomen* (*aa*). Extensive genetic and molecular studies have been performed upon the *aa* syndrome. Genetically, this syndrome is associated with two necessary genetic components which are found on the X chromosome less than 1 map unit apart. Molecularly, the syndrome also depends upon two necessary components. First, a majority of the 28S ribosomal genes must have a 5 kb insertion into their coding region. The ribosomal genes are found in a tandem cluster of about 200 copies on the X chromosome. Second, during the formation of polytene tissue (such as fat bodies), there must be no preferential replication of non-inserted over inserted 28S genes. Many X chromosomes have the ability to replicate preferentially the non-inserted sequences both intra- and interchromosomally. If such a preferential replication occurs, *aa* is suppressed. In addition to these major and necessary X-linked elements, the expression of *aa* can be modified by genes located on the autosomes. Although the major genetic elements for *aa* are on the X chromosome, this syndrome is also separable through genetic recombination from the sexual isolation and male sterility X-linked elements.

Notice that all three isolating barriers have one or more major genes on the X chromosome, but most autosomes are not involved or have only minor modifiers. Given that the X chromosome constitutes only about 20% of the *mercatorum* genome, this preferential involvement of X-linked genetic systems is indicative of the increased sensitivity of the X chromosome to founder events over that of the autosomes. An alternative explanation is that genes involved with isolating mechanisms are clustered upon the X chromosome. This later explanation seems unlikely, since a broader review of the genetic basis of isolating mechanisms in the genus *Drosophila* shows extensive autosomal involvement.

Natural speciation events

There is evidence for this same pattern of preferential involvement of X-linked genetic systems in natural speciation events associated with founder effects. Founder-induced speciation is probably the dominant mode of speciation among the Hawaiian *Drosophila*. Two of these Hawaiian species are *D. silvestris* and *D. heteroneura*. These species are morphologically distinct and are reproductively isolated by premating barriers that can be broken down in the laboratory. Since no postmating barriers are present, it is possible to perform genetic analyses of the traits differentiating these two species. One of the more dramatic differences is in head shape. *Drosophila heteroneura* has a hammer head, whereas *D. silvestris* has a round head. Head shape is controlled by a major gene or genes on the X chromosome epistatically interacting with several minor autosomal modifiers. The species also differ in carina and groove (face) colour, with *D. heteroneura* being yellow and *D. silvestris* black. Both aspects of face colour are controlled by an X-linked locus interacting with one autosomal gene. Two different aspects of wing pattern are each controlled by interactions between an X-linked and autosomal locus. Differences in the pigmentation of the mesopleurae are due to an autosomal locus. Hence, five of the six characters analyzed involve major genes on the X chromosome. Thus, both experimental and natural founder populations display a preferential involvement of X-linked loci for their evolved differences.

Further evidence for the preferential sensitivity of X-linked genetic systems to founder events can be obtained by looking at the pattern of fixation of chromosomal inversions among the Hawaiian picture-winged *Drosophila*. Of 127 inversions fixed in 103 species, 59 (46%) are on the X chromosome despite the fact that there are four autosomes of about equal size. One might explain this by assuming that the X chromosome has a greater rate of inversion mutation, but this seems unlikely. If fixation rate is merely a consequence of mutation rate, the proportion of fixed to polymorphic inversions should be the same across all chromosomes. The contingency chi-square test of this hypothesis is 23.19 with four degrees of freedom, a result that is significant at the 0.00001 level of significance. Excluding the X chromosome from the analysis, the resulting contingency chi-square is 0.66 with three degrees of freedom. Hence, the autosomal inversions are being fixed at a rate that is proportional to their polymorphism rates, but the X chromosomal inversions are being fixed at a rate that is far in excess of their polymorphic rates. More specifically, the data indicate that X chromosomal inversions are being fixed at a rate 4.9 times that of autosomes. As a further check on the inference that these differences can be attributed to founder events rather than some peculiar property of the X chromosome, also presents the comparable data set on inversion evolution in the *repleta* group of *Drosophila* as summarized in Wasserman (1982). This is a speciose continental group for which there is no evidence for extensive founder-induced speciation. In great contrast to the Hawaiian *Drosophila*, only 10% of the fixed inversions are on the X chromosome, and the contingency chi-square between fixed versus polymorphic inversions as a function of chromosomal location is 3.62 with four degrees of freedom. Hence, in the *repleta* group, the rate of inversion fixation is proportional to rate of inversion polymorphism, and there is no discrepancy between X versus autosomal inversions. The contrast between Hawaiian versus *repleta* group *Drosophila* provides strong support for the theoretical prediction that X-linked genetic systems are far more sensitive to founder events than autosomal genetic systems.

Mitochondria DNA versus Nuclear Genetic Systems

The theory outlined above predicts that mtDNA should be more sensitive to single-female founder events than nuclear systems. To test this prediction, DeSalle (1984) performed an extensive study of mtDNA restriction site evolution in two lineages of Hawaiian *Drosophila*, the alpha and beta lineages of the *planitibia* subgroup. To understand why these two lineages where chosen, it is first necessary to review some pertinent facts concerning the geological history of the Hawaiian Islands. Most of the major islands of the Hawaiian chain are separated by deep ocean channels. The fluctuating sea levels that occurred during the Pleistocene could never have connected these major islands. Hence, the traditional arguments for single female founder events are applicable to instances of speciation involving transfers between these major islands. Most speciation in the beta lineage falls into this category. However, Maui, Molokai and Lanai (known collectively as Maui Nui) are islands that were connected to one another at various times in the past as a function of fluctuating sea levels. Hence, the *Drosophila* on these islands, which includes the alpha lineage, were much more likely to be separated by subdivision of large populations rather than by founder events.

Behavioural studies provide further evidence that founder events were not important in speciation processes occurring on Maui Nui. Kaneshiro (1976) discovered a pattern of mating asymmetry between ancestor and descendant species that exists among many Hawaiian *Drosophila*. Basically, derived females will mate with ancestral males, but ancestral females will not readily mate with derived males. This pattern has also been observed in non-Hawaiian *Drosophila* for which biogeographical evidence indicates founder-induced speciation. In addition, this mating asymmetry evolves in experimental founder events involving both Hawaiian and non-Hawaiian *Drosophila*. This pattern is not observed in contrasts between continental *Drosophila* species for which the biogeographical evidence indicates founder events to be unlikely and is not present in *Drosophila* speciation experiments that do not involve founder effects. Consequently, this pattern of mating asymmetry is one that appears to be specifically associated with founder-induced speciation in *Drosophila* for the theoretical reasons why this specificity is expected). Most Hawaiian *Drosophila*, including the beta lineage, display this asymmetry. The only exceptions discovered so far involve species from the Maui Nui complex, thereby implying that speciation on Maui Nui does not involve founder effects.

To see if mtDNA is more sensitive to founder events than nuclear genetic systems, comparisons were made between DeSalle's genetic distances on mtDNA using Ewens' (1983) 'p' statistic with Roger's genetic distances calculated by Johnson et al. (1975) on isozyme data. The Ewens' and Roger's distances are not directly comparable, so the absolute values of the distances are not important, but rather the relative pattern within and between lineages.

The pattern of distances within and between lineages can be summarized by taking the appropriate averages. For the isozyme data, the average distances within lineages are 0.10 and 0.27 for alpha and beta respectively, and the average intralineage distance is 0.62. Hence, in terms of the nuclear genetic distances, the interlineage distance is considerably larger than intralineage distances. Moreover, as Carson (1976) has shown by overlaying these genetic distances on the known geological time scale of these islands, the distances in flies from both lineages can be explained by the same rate of evolution. In contrast, the average intralineage p-distances for the mtDNA are 0.028 and 0.053 for alpha and beta respectively, with the interlineage distance being 0.059. This pattern differs dramatically from that observed for the isozymes. Note that the interlineage distance is only slightly larger than the within beta distances. This pattern implies that the beta lineage has a much faster rate of mtDNA evolution than the alpha.

The hypothesis that the beta lineage is evolving more rapidly than the alpha lineage for mtDNA can be tested directly using the statistical procedure outlined by Templeton (1983b, 1986b). This procedure uses the character-state data rather than distances. A great advantage of character-state data over distance data is that it is often possible to reconstruct the ancestral intermediates between the taxa being compared. This property has long been used in constructing intermediate states in the inversion phylogenies of *Drosophila*, and the logic is identical in reconstructing intermediate states of restriction sites. This is an important property because it allows one to estimate the number of mutations accumulated along any branch segment with an ancestral-state estimation procedure that does not assume the molecular clock a *priori*.

A second desirable property of this test is that it is non-parametric. Hence, no assumptions are made about the underlying sampling model induced by the evolutionary process. This is important because we simply do not know at this time what the appropriate sampling model ought to be; yet, the conclusions based on parametric tests of the molecular clock are very sensitive to the sampling model assumed. However, recent work indicates that the molecular clock does not hold.

The details of this test are given in Templeton (1983b, 1986b), so they will not be repeated here. Nevertheless, some of the limitations of this approach should be mentioned. First, the test requires a knowledge of the topology of the phylogenetic tree. Fortunately, Templeton (1983b, 1986b) gives statistical criteria for judging how well a topology has been estimated. If more than one topology cannot be statistically distinguished, it is best to test the molecular clock under each topology to check for the robustness of the conclusions with respect to the ambiguities regarding the true topology. DeSalle (1984) used these statistics on the Hawaiian species being considered, and concluded that there were two statistically indistinguishable phylogenies.

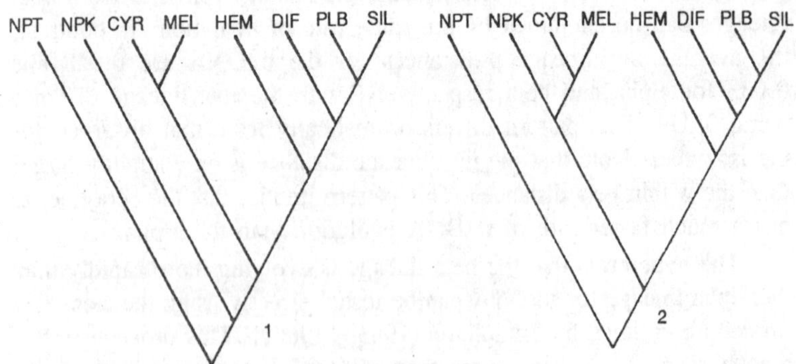

Fig. 5.1. Alternative phylogenies for the Hawaiian Drosophila species under consideration.

Second, given a topology, the ancestral states themselves may not have been estimated in an unambiguous fashion. The reason for this is that some alternative ways of allocating mutational events or even the number of events are basically equally probable. To prevent this estimation ambiguity from eroding the power of the test, it is important to only test data sets for which convergences of any sort are rare. Fortunately, DeSalle's data set falls into this category.

The pattern is very clear cut; the beta lineage is evolving at a greatly accelerated rate compared to the alpha lineage, Hence, founder events do seem to accelerate the rate of mtDNA evolution over that of nuclear genetic systems. When these results are coupled with the earlier results concerning X-linked genetic systems, it is clear that the mode of cladogenesis interacts with the nature of the genetic system to determine rate of evolution.

Anagenesis and Genetic Systems

The debate concerning punctuated equilibrium has focused much attention on the mode of speciation as a primary determinant of rate of morphological evolution. In contrast, most models of molecular evolution have ignored the details of speciation mechanisms and explained rates of molecular evolution solely in terms of continuously operating anagenetic mechanisms, primarily genetic drift. As shown by the contrast of the alpha and beta lineages given above, this disregard for speciation mechanism by molecular evolutionists is clearly not justified. Templeton et al. (1981) give additional examples of the importance of speciation in determining rates of molecular evolution, and Gillespie (1984) has recently performed a statistical analysis that indicates an episodic pattern of molecular evolution. However, anagenesis is still important in determining evolutionary rates, as will now be shown.

The importance of anagenetic trends as a determinant of rates of molecular evolution will be illustrated by examining humans versus African apes. Given the reasonable assumption that genetic drift is the primary determinant of rate of fixation of most molecular variants, there are at least two anagenetic trends that should cause rate heterogeneity between humans and African apes. First, one of the important features of human evolution has been a dramatic decline in the rate of development and an attendant increase in generation time over the past three million years. The rate of evolution through genetic drift is proportional to generation time unless one makes the assumption that mutation rates are proportional to absolute time. This assumption is not consistent with the empirical evidence among eukaryotes. For example, estimated mutation rates in *Drosophila*, mice, corn and humans are generally within an order of magnitude of one another even though the generation lengths of these organisms differ by three orders of magnitude. Although mutation rates are hard to estimate accurately, it is difficult to reconcile such grossly discrepant results with the idea that mutation rates are constant in absolute time in higher eukaryotes. Indeed, Kimura (1979), long one of the leading advocates of mutation rate constancy in absolute time, has abandoned that position and now models his version of the neutral theory in terms of mutation rates that are constant on a per generation basis. Hence, the straightforward prediction of the anagenetic trend to longer generation times in humans is that the rate of molecular evolution in humans should be slower than that of African apes. Moreover, this slowdown should affect all genetic systems.

Another trend in the human lineage, but not in the African apes, has been a consistently expanding population size starting at the very least with *Homo erectus* which had a fossil distribution covering the entire Old World and not just Africa. Nei (1976) has shown that expanding population size decelerates the rate of molecular evolution under absolute neutrality, and Kimura (1979) has shown that the deceleration is even more severe when one includes mutations with very small selection coefficients. Once again, this slowdown should affect all genetic systems in which molecular evolution proceeds by random fixation of neutral or nearly neutral mutations.

These predictions can be tested by comparing the rates of molecular evolution in humans versus African apes for a number of genetic systems. First, Templeton (1983b, 1986b) has already tested the rate of mtDNA evolution in these primates and has shown that there is a statistically significant slowdown in the rate of mtDNA evolution in humans relative to African apes. This slowdown also appears to exist for nuclear genetic systems as well. For example, Marks (1982) concluded that humans have a slower rate of karyotypic evolution than the African apes. More recently, Sibley and Ahlquist (1984) performed extensive studies on nuclear DNA-DNA hybridization between these species. Testing the molecular-clock hypothesis is far more difficult and less clean-cut with distance data than with character-state data and, unfortunately, the DNA-DNA hybridization data is inherently a distance measure. Nevertheless, inferences can be made concerning the clock, but they must be made carefully rather than from a casual inspection of the data.

For example, Sibley and Ahlquist (1984) argued that their data is consistent with the clock because it satisfied the triangular inequality. They then quoted Farris (1981) to the effect that this metricity implies 'clocklike' behaviour. However, what Farris (1981) really said was that metricity is a necessary, but not sufficient, condition for clocklike behaviour. Commenting on Sibley and Ahlquist's quotation of Farris, Farris and Kluge corrected this error and pointed out that 'a distance may be metric, yet show strong variation in rates of evolution'.

A better way of testing the molecular-clock hypothesis is to contrast the distances between two closely related species with a third, more distantly related, outgroup species. This is known as the relative rate test. The relative rate test, however, is biased in favour of the clock unless one can correct for the covariances of the distances or estimate the ancestral distance in a manner that is not biased towards the clock. Unfortunately, the covariances for DNA-DNA hybridization distances cannot be calculated unless one invokes a parametric model of unknown validity. The second course of action, estimating the ancestral distance states, is flawed by the fact that distance estimation algorithms are biased in favour of, or explicity assume, the clock. For example, Nei et al. (1985) recently gave formulae for estimating the position and variance of branch points for various types of distance data. However, their equations are only valid if the molecular clock is exactly true, so their estimation procedure cannot be used to test the clock.

Although it is difficult to test the clock hypothesis with the Sibley and Ahlquist data set, it can easily be shown that these data contain internal inconsistencies which cannot be reconciled with the hypothesis of rate constancy. For example, the distance between humans and chimps is significantly smaller (at the 0.1% level) than that between humans and gorillas, but in contrast the genetic distance between gorillas and chimps is significantly smaller (at the 1% level) than that between humans and gorillas. These conclusions are based on t-tests that were uncorrected for the covariance between distances, and hence these tests are strongly biased in favour of the clock. Despite this bias, these significant t-test results cannot be reconciled with the clock. If the clock were true, the first significant test would imply that humans and chimpanzees form a clade, but the second test implies that gorillas and chimpanzees form a clade. Since there can be only one phylogenetic topology, the molecular clock obviously cannot hold for all these species. In a survey of the available molecular data, Templeton (1985) showed that there is one and only one branching order that is statistically compatible with all the molecular data; namely, the tree that clusters the African apes together as a clade. As further discussed by Templeton (1985), the pattern of internal inconsistencies found in the data of Sibley and Ahlquist (1984) can be readily explained by a slowdown in the human lineage under this topology because, although humans are the most distantly related in a topological sense, the human slowdown in rate of molecular evolution could yield very similar expected distances among all three species.

Thus, both the nuclear and mtDNA genetic systems appear to have been slowed down in humans relative to the African apes. This pattern supports the predictions discussed above based on the well-documented anagenetic trends in the human lineage of increasing population size and generation times.

Discussion

The examples given in this chapter illustrate that rates of molecular evolution are not constant, but vary as a function of both cladogenesis and anagenesis. This variation in rates of molecular evolution does not mean that molecular data cannot be used for making inferences about times of divergence. The molecular data can still yield much information concerning divergence times, and indeed, the identification of specific heterogeneities can allow times of divergence to be estimated in an even more precise manner. These estimated numbers of mutations will be used as a distance measure. This distance can be calibrated in absolute time by noting that the divergence between *D. planitibia* and *D. silvestris* involved a founder event on the Hualalai Volcano on the Island of Hawaii, which is about 400,000 years old. Assuming this founder event occurred shortly after the formation of Hualalai, the *silvestris/planitibia* divergence is set equal to 400,000 years. Another potential calibration point is the divergence between the alpha and beta lineages. The chromosomal and biogeographical data indicate that these lineages diverged on West Maui, which is 1.3 million years old. However, for now, only the *planitibia/silvestris* calibration will be used.

Using this calibration with no correction for the rate heterogeneity indicated one obtains the divergence times. The divergence times of *differens* and *hemipeza* are consistent with the available biogeographical data; a result that is not too surprising given the fact that the clock holds within the lineages. However, the divergence between the alpha and beta sublineages is estimated as 835,000 years, a figure that is much too small. Moreover, the estimated times of divergence within the alpha lineage are low (74,000 years for *crytoloma* and *melanocephala*, and 185 000 years for *neoperkensi*).

Because highly significant rate heterogeneity was detected between the alpha and beta lineages, but not within them, the divergence times were next estimated by using the original calibration on the *silvestris/planitibia* node only within the beta sublineage. The rate of mtDNA evolution within the alpha lineage was regarded as being 3.23 times slower than that within the beta. This figure was obtained by dividing the average number of mutations on the branches leading to beta species from the alpha-beta node by the average number of mutations on the branches leading to alpha species from this node. Using the calibration corrected for rate heterogeneity produces the estimated divergence times. These estimated times make much more sense in the light of the known biogeographical data. For example, the alpha-beta split is now estimated to have occurred at 1.27 million years; a figure that is in excellent agreement with the 1.3 million year age of West Maui. Recall also that the behavioural and biogeographical evidence mentioned above implies that the alpha-lineage flies did not speciate through interisland founder events, but rather through the more traditional erection of geographical barriers between Maui Nui subpopulations. Since the *neoperkinsi* strain currently lives on Molakai (1.5 to 1.8 million years) and the *cyrtoloma* and *melanocephala* strains on East Maui, the speciation events separating these groups could only have occurred after the formation of East Maui if this interpretation of their speciation pattern is correct. Note that the estimated divergence times in part B are consistent with this interpretation, since it indicates that the first split in these alpha species occurred shortly after the formation of East Maui.

The results reported in this chapter not only reveal that cladogenesis and anagenesis can affect the rate of molecular evolution, but in addition the results show that sometimes the impact on evolutionary rates is uniform on all genetic systems within a species (as shown by the human example), but in other cases the impact is quite heterogeneous (the *Drosophila* examples). Hence, different genetic systems imbedded within a single species are differentially sensitive to some of the historical contraints that shaped or even created that species. By simultaneously studying several different genetic systems, patterns emerge that can be used to make inferences about the history of the species being examined that could not be made just by studying each genetic system in isolation. For example, Charlesworth et al. (1982)

argued that founder events were unlikely in Hawaiian *Drosophila* because they have maintained high levels of isozyme heterozygosity. However, Charlesworth et al. (1982) noted that this observation does not exclude founder events followed by a population flush because a depletion of isozyme heterozygosity is only expected if the population 'bottleneck' is 'prolonged over several generations'. Consequently, the isozyme data, considered in isolation, are consistent with two hypotheses: one, no founder or bottleneck effects occurred; or two, founder events occurred followed by large increases in population size. This ambiguity in interpretation is clearly resolved when the mtDNA data is overlaid on the isozyme data, for now the no-founder hypothesis can be rejected on the basis of the dramatic rate of increase of mtDNA. Consequently, a simultaneous examination of evolutionary rates among various genetic systems offers a potentially powerful tool for making inferences about the cladogenetic and anagenetic events that shaped the evolution of the species we observe today.

6

HOMEOSTATIC EVOLUTION

The tendency to increase in size is, in a sense, the most fundamental feature of living things. We are all familiar with it, and take it for granted as a 'natural' process. Yet even slight study will show that growth is very far from simple and that we do not properly understand it. Moreover, it is very 'unnatural', since it is unusual among physical phenomena. It involves incorporation of less-ordered material into the highly ordered living systems, contrary to the general tendency for increase of entropy. It is indeed worth while to study growth closely, and we may expect to have surprises when we do so. There are many ambiguities even in attempting to define growth. It has been called 'the net balance of mass produced and retained over mass destroyed and otherwise lost'. But is all increase of mass to be called growth? Is addition of storage fat or carbohydrate as much 'growth' as addition of protein? And what about the addition of water? How are we to define the mass of any organism? It fluctuates from moment to moment.

Features other than mass may serve to define growth. Height would be regarded as a good criterion by many children. Increase in cell number is obviously linked in some way with the concept of growth, but individual cells may grow, so the increase of number of cells alone cannot define the process of growing. From this point of view we might distinguish between (a) the addition of material to cells, (b) increase in their number, and (c) increase of intercellular material.

It would be satisfactory to be able to put forward some concept that would unify these various aspects of growth. The differences between them are semantically puzzling but trivial. A true measure of growth might be related to the rate of increase of order that is implied by the incorporation of non-living into living matter. The increase of DNA in a population might serve as the basis for an assessment of the degree of order, which would evade these difficulties about fat, water, etc. This statistic (amount of DNA) is indeed very useful for some purposes, but we may also have to be content to measure growth by whichever of the paprameters discussed above is most readily available to us. If a general phrase is wanted we can perhaps say that growth is 'the addition of material to that which is already organized into a living pattern'.

We shall be concerned here not so much with definitions or even with finding quantitative expressions for growth (though this will follow). It is more important to try to see the relation of growth to the processes of self maintenance. In fact we begin by asking why should there be growth in order to ensure homeostasis? This may seem a very large question. And indeed so it is, for it touches the very nature of living organization. A system of steady-state interchange is able to continue only because it is provided with 'instructions' that ensure correct responses in the situations that are likely to arise. To put it in another way, an organism an continue to live only because it has been provided with the necessary organs. The instructions thus constitute a system for forecasting the situations and risks that the creature is likely to meet. But by the very nature of the randomness of events, something will happen sooner or later that destroys part of, and eventually the whole, organism. Continued survival in a steady state by any one given set of materials is therefore impossible. Survival of the organization can, however, be ensured by a suitably arranged system of replacement. This is, therefore, the 'function' of growth. It operates at a whole series of levels, from the continuous replacement by turnover within cells, to continuous replacement of cells, to replacement of whole individuals by others, in the processes we ordinarily call reproduction, and in the replacement of one species by another in evolution. All these are a single series of phenomena by which the power of replication and of incorporation into the living system serves to ensure the maintenance of life.

Exponential Growth

One of the most unexpected findings for the layman is that growth, if it is not restricted, continues without limit. What grows is itself capable of growing. This is well illustrated by bacteria. When bacterial cells are isolated in a culture medium there is first a lag period and

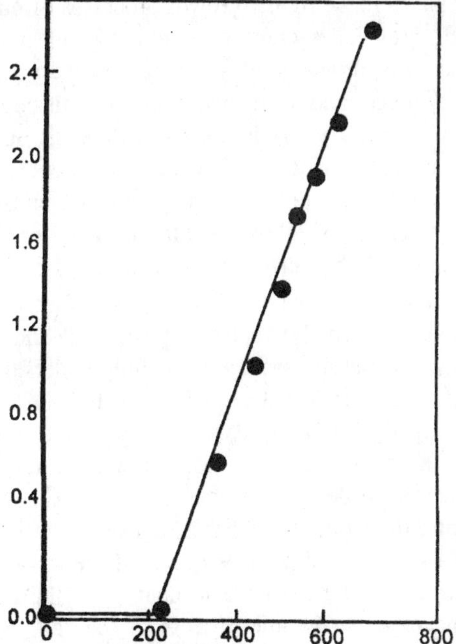

Fig. 6.1. Growth of bacterial cells in culture afer explanation.

then exponential growth begins. Each bacterium divides about once every 30 to 40 minutes and the products then 'grow' until they reach the size of the original. This process continues until the supply of nutrients is exhausted, after which special conditions set in, which do not interest us here. If the culture medium is renewed, growth continues indefinitely. Use is made of this for continuous cultivation methods. If all the growing organisms were to be supplied with the required nutrients there would have to be an exponential increase of the size of the vessel and the amount of inflow. This is obviously not possible, but if part of the volume of multiplied micro-organisms is withdrawn at the same rate as nutrient medium is supplied, then a steady state can be maintained in which all the materials are utilized. More penetrating observations have shown that we must think more carefully about continuous growth. Does each individual grow at an increasing rate per unit of mass, as would be required if what grows is capable of growing?

Mitchison studied the growth of single cells of yeast, using interference microscopy to determine the 'dry weight' of single cells continuously. He kept the yeast in a sterile medium so that there was no question of interference by the growth processes of other organisms. Each yeast cell grew at a constant rate independent of mass, until late in the process of division, which occurred every four hours, and then the rate doubled and remained the same throughout the next generation. The rate of synthesis therefore depends upon a group of components, perhaps the ribosomes, which may double their numbers before cell division. The whole process of division must obviously be related to the replication of DNA and this, in many organisms, occurs at about the middle of the period between divisions.

The determination of the DNA cycle probably depends upon a complex relation between the replicating material and the associated polymerase enzymes, but the system is not fully understood. The whole question of the basic cycles that regualte cell division is still obscure. The cell does not divide because it has reached a certain size. By various means the growth can be stopped prematurely, but division will take place, though delayed. Addition of material is therefore one factor influencing division, but not the only one. Probably replication of the DNA is a second, and perhaps cycles of the RNA and ribosomes a third. A fourth may be a rhythmic change in the energy-yielding mechanisms and other enzymatic process of the cell, perhaps related to the cycle of reducing sulphur-containing compounds (e.g. glutathione).

In examining the fundamental process of control of increase we find, therefore, many facts that are interesting and sometimes paradoxical, but difficult to reduce to any single system of control. Life may perhaps once have been simple, but today even the lower organisms maintain their balance through a web of interacting control mechanisms. We have, however, the important conclusion that the general condition is to continue growth indefinitely, at least in these simple organisms. The situation with which we as humans are familiar, namely, decline and cessation of growth, followed later by death, thus appears as a limitation of the general condition. Our thesis will revolve around the concept that our life cycle indeed includes special sets of limitations upon continuous growth, serving to ensure that the population as a whole can continue to survive by means of a particular way of life.

Differentiation and Growth Control

Perhaps it should not surprise us to find that continuous growth is the general condition. Since continuous replacement is the absolutely necessary condition of self-maintenance we should expect to find it everywhere in living processes. The growth must of course be regulated, especially in the higher organisms, where survival depends upon the presence of a most complicated system of tissues, differentiated for particular functions. Differentiation is in this sense the inverse of growth; indeed, it often follows growth. We shall expect to find, then, that the study of growth consists largely in the study of the control and limitation of growth, especially in the higher organisms.

Intermittent Growth

When we look more closely, it thus appears that even in bacteria growth is not continuous. Addition to the mass of a whole culture masks the cycles in the growth of each individual. Replication of the DNA occurs at set intervals, and something is known of the means for regulating this. In the absence of limitation by lack of raw materials, it seems that the single strand of DNA in each bacterium is replicated by the attendant polymerases at intervals that are not closely related to the total increase in mass. The factor that regulates this cycle thus serves to control the whole growth potential of the species. Presumably, however, this cycle is itself genetically controlled. This question of the power of increase of a set of cells comes very close to the heart of all biological problems.

Unfortunately, we have only a glimpse of it in bacteria and viruses and known almost nothing of it in higher animals and plants. It is difficult to conceive of the process of growth as a wholly continuous phenomenon, for the simple reason that organisms themselves are essentially heterogeneous. We can think about the continuous synthesis, say, of a given protein. But the more interesting proteins do not simply accumulate, they associate with other substances to form membranes, organelles, fibres, and so on. When any one of these has been formed the continuity is in a sense broken as the building of another organelle is begun. The point is not trivial, for organisms are assemblies of such units. Higher organism are assemblies of assemblies of them (cells, tissues, colonies, populations). Since the components of such systems do not increase together and continuously we have to think of them separately and individually. The origin and fate of each mitochondrion or man is therefore inevitably a large part of the study of biology. Investigation of the cycles of their multiplication and growth may reveal some general principles, but is also likely to show that there is a *great diversity of rules of growth.*

Turnover and Growth

A further complication is that addition to living systems is not a one-way traffic. We must also include the subtraction that is involved in breakdown, and then compute the growth as the net gain. It seems so obvious that growth is the resultant of anabolism and catabolism that we do not pause to ask 'Why should there be breakdown at all?' More practically, we may seek an answer by looking at lower organisms to see whether turnover is always present. Perhaps it is necessary feature for the maintenance of a steady state. Viruses cannot be said to show turnover, but they are not capable of independent self-maintenance.

In bacteria there is turnover, btu at a simple level than can hardly be distinguished from reproduction. In the more familiar situation of multicellular organisms, it has been clear since the time of Bizzozero (1894) that some parts of the body become changed much faster than others. Whether we are considering cells alone or their components (fibres, organelles, etc.) it is convenient to think of three types.

1. Labile populations, multiplying throughout life (vegetative intermitotic cells of Cowdry (1952)).
2. Stable populations, usually multiplying only during development, but capable of doing so in the adult after damage (reversible postmitotics).
3. Permanent populations, never multiplying once formed (fixed postmitotics).

As applied to cell populations in mammals this would produce the following classification:

Labile (renewing) populations

Epidermis and its derivatives

Endodermal epithelium

Transitional epithelium of ducts

Blood-forming tissues (in part)

Endometrium of uterus

Germ cells of the testis

Stable populations (able to expand when necessary)

Liver

Kidney

Exocrine glands

Lens

Connective tissues

Skeletal tissues (in part)

Permanent populations

Striated muscle
Neurons
Germs cells of the ovary

Turnover and the Anticipation of Risks

In spite of our considerable understanding of the functions of the various tissues in the economy of the body it is not always clear why some tissues should remain intact and others be regularly replaced. Moreover, little is known about the important question of what determines the rate of production of the labile tissues or stops the stable ones from further expansion. The products of the epidermal line are ultimately keratinizeid. Thus removed from metabolic sustenance, they are subject to the cruder forms of wear and must be replaced. This simple situation may serve as an example of the principles that we may search for. The body as a self-maintaining system inevitably suffers the physical hazards imposed by a more or less tempestuous environment. Some of these risks are of simple physical nature (like the wearing away of the skin), others are chemical changes or fluctuations of temperature. No portion of the body can be regarded as permanently immune from risks. However strong the bones, in some individuals one or another of them will ultimately be broken. To make them still stronger would involve not only collecting much material but, worse, carrying it around.

The tortoise provides himself with a house of inert material that is hard to break. Within it he lives longer than, perhaps, any other creature, but tortoises are not among the most widespread animals and have not been able to colonize may habitats, for example the trees or the air. Such massive over-insurance has not proved to be the ideal policy for life. In general, the best protection seems to be provided, not by excessively strong or inert materials, but by a system of turnover that allows for replacement.

Cell replacement may proceed continuously, as in the skin, or be called into play when needed, as in the repair processes of the connective tissues or skeleton. In this sense the continuous change of the renewing populations is an aspect of homeostasis. It provides a permanent insurance against the inevitable damage to the components that is imposed by the environment. That this is likely to be an efficient policy is clear, for example, in the skin, whose material is directly subject to wear. It is also easy to understand for the lining of the alimentary canal, which must deal with all the chemical influences that are introduced with the food, including the products of the bacteria that thrive there.

In other tissues it is less clear why continuous cell-replacement is needed. There is no obvious need for rapid renewal of the red corpuscles of the blood. Possibly the limiting factor here is the energy needed to pump them round the body. It is cheaper to pump sacks of almost pure haemoglobin then to destroy them after a while, and replace them from a factory, than to pump also the instructions, machinery, and materials for their renewal. Incidentally, birds find the latter course better. Their red cells, though they are still end cells, have nuclei, and this enables them to manage with a small factory and so to lighten their bones with air sacks, while haemopoiesis on a much reduced scale proceeds in the liver, spleen, and elsewhere.

Cell replacement is obviously not the only possible means of renewal. In some tissues replacement is probably undesirable for functional reasons. Thus the activities of the nervous system depend upon an elaborate system of connectivity between the cells. To renew this would need complicated instructions and in fact regeneration can be achieved only in simpler animals and simpler parts of the nervous system. No effective regeneration takes place in the brains of mammals (or of cephalopods). Where the brains contain memories, capable of holding records of the experiences of the individual, it may obviously be difficult to provide also instructions to replicate these records if cells are destroyed. Correspondingly the neurons lose all power of cell division. They do not, however, lose their powers of synthesis. On the contrary, protein turnover is especially active in the nervous system.

Nerve cells are characterized by a large amount of ribonucleic acid in their cytoplasm—the basophilic Nissl bodies. Probably there is a continuous flow of material down the nerve fibres from the cell bodies. What happens to it at the peripheral end is uncertain. It is clear, therefore, that these static post-mitotic' cells are not really different from the rest. Indeed it may be that they are provided with especially active mechanisms for self-renewal. For, unlike the other populations, they cannot be replaced if damaged, and must therefore continuously repair themselves. This presumably applies to striated muscle as well as to nerve.

The very large number of nuclei in each syncytial muscle fibre suggest the operation of continuous processes of replacement (though myosin itself turns over little it at all). It becomes clear, therefore, that the phenomenon of growth, that is to say, synthesis of new material, is much more than the production of a new individual, as we commonly regard it. Growth is one of the fundamental features of homeostasis; growth provides the means of replacement of what will inevitably sooner or later be damaged—a guarantee, so far as possible ,against the risks of destruction.

The growth with which we rare familiar, that of the child, after the act of reproduction, we shall find to be the special from of this guarantee, since it allows survival *in face of almost all risks,* a promise for the future that would seem impossible to give, but which has so far served to ensure life for 3000 million years or more. For the purpose of a general study of growth, then, our conclusion is that addition to the existing mass of living material is a fundamental property, presumably existing since life began. The numerous special forms of life that exist today have been arrived at partly by limiting this general power of growth. Thus all the higher organisms are colonies of different types of cells, serving different functions in the homeostatic system. the number and position of each type is regulated to produce a smoothly working whole. The study of growth thus becomes, in effect, the study of what limits growth in each tissue, during development and in the adult.

Repair of the Individual

Maintenance, Repair, Replacement, and Regeneration

These are a series of similar processes, linked both in their significance for continued maintenance and through the methods by which they operate. All are processes by which the individual life is enabled to persist in spite of actual wear or damage. As we have seen in the last chapter, in most tissues there is continual turnover and in some the cells are regularly replaced. These maintenance operations anticipate the damage or destruction of the parts in question, and they are found particularly in such vulnerable areas as the outer skin and lining of the gut. Other parts of the body, though liable to damage, may suffer it only 'accidentally', that is, occasionally or not at all. For these the genetic system provides processes of healing, repair and replacement that are set into action only when called for as a result of traum.

'Regeneration' is usually kept as the name of a similar process but on a larger scale, leading to the production of a whole new part such as a limb. These powers are therefore, as it were, a second line of defence when the forecasting mechanisms that avoid damage have failed. For example, the mechanisms of heredity and development provide a mammal with bones of strength adequate to meet most of the stresses likely to fall upon them. If greater stresses are met and a bone breaks, then a specific mechanism of regeneration is available to mend it. All these processes involve growth, and are linked with the daily turnover within the tissues. They also involve cell migration and differentiation, and in this are linked with normal development from the egg.

All of these repair operations that we usually separate conceptually for convenience under the headings given above form a series running from those more closely related to turnover to those that partake more nearly of reproduction and development. Each uses specific mechanisms to contribute to the homeostasis of the system. In this sense all are 'normal' processes. They differ in the regularity and timing of the wear or accidental events that they are designed to correct. Processes of 'maintenance' are designed to meet the damage that occurs in certain tissues regularly every day, for example in the skin of man, or the incisor teeth of a rodent. 'Repair', 'replacement,' and 'regeneration' are processes that come into operation only intermitently, when the organism has suffered one of the various hazards that are likely to befall it. In this sense they approach nearer to the mechanism of reproduction, which is the act that initiates the replacement of the whole creature before eventually something happens by which it is destroyed. The earlier categories, maintenance and repair, are thus more directly concerned with minor, recurrent, and temporary damages; the last two, especially regeneration, constitute major acts of differentiation, approaching reproduction.

Indeed, in some organisms it is hard to distinguish between regeneration and reproduction by budding (e.g. sea squirts (tunicates) or tapeworms). Since the processes of repair, replacement, and regeneration are characteristically brought into operation after accidents or unusual events in the life cycle, they are obviously of particular concern for the medical man. He is, for many people, the specialist who is called in to help with the results of such untoward stresses, rather than the guardian of longer-term homeostasis (though he may well play a great part in this to, and will perhaps do so even more in the future).

In order to be able to assist the repair processes of the body effectively, the doctor must obviously know about their means of operation. It is to some extent the function of pathology to provide this understanding. Here, however, we meet with the complication that, as a prelude to the study of repair, the pathologist also investigates the processes of damage themselves, especially as inflicted by invasive agents such as bacteria and viruses. We shall not attempt to study pathological conditions here but shall try to investigate the nature of the healing processes as exemplified in a few examples in which they have been set into action by controlled surgical experiment. Elementary principles can be illustrated in this way, but while the reactions to particular pathogens are alike in the early stages, later the picture may change specially for each disease condition.

Maintenance

Of the four categories that we are considering this approaches most closely to the normal turnover processes. The concept of maintenance, as borrowed from engineering and household management, implies the repair of wear and tear that is normal in the sense perhaps of 'expected'. We have to paint our houses and renew electric light bulbs or motor-car tyres or brake-linings, because the designers of these items were not able to make them more permanent. A similar distinction can be made for the body, though it is not perhaps altogether appropriate. The body has no permanent components (unless it be the instruction in the DNA), but it has some parts that turn over very little (e.g. the collagen of tendons, actomyosin of muscles, or antibodies), while others turn over so efficiently that we do not ordinarily think of them as suffering wear (e.g. the continual replacement of red cells or intestinal epithelia). Perhaps, however, this is indeed a wrong attitude and we should recognize that it is exactly because of the inevitable molecular wear and degradation that the turnover of these tissues is so active.

In addition, of course, it allows the continual adjustment of the level of their functioning by adaptation. The conceptions of maintenance and repair should be extended for living organisms to the molecular level. Some enzyme molecules persist only for a few minutes. It is not clear in what sense that is due to wear, but in muscle there is an increased turnover after activity. The replacement of the surface of the skin or nails from deeper layers is more obviously an act of maintenance. We find it easy to conceptualize this process because we can literally *see* it at work on our own bodies.

There is no reason to doubt that with sufficient insight we should equally recognize that the replacement of our red corpuscles and all the other replacing tissues shows the operation of specific mechanisms under the control of the DNA, ensuring maintenance of the integrity of the system in spite of wear. As a steady-state open system, the organism is continually subject to stresses resulting from the tendency to merge with its surroundings. These knocks provide the stimuli for the operation of the synthetic growth processes, which are made possible by the replicative powers of the instructional molecules within. Each tissue is thus maintained at the appropriate level of functioning by a *double dependence*, on stresses from without and synthetic forces from within. The problem for the student of growth is to learn to identify and measure these forces so that he can estimate the probability of successful continuance and show how to increase it by appropriate action.

To take a simple example, in managing the healing of bone the physician needs to know how the effect of mechanical stress will influence the process at each stage (specifically, when and how much to let the patient with a broken leg walk.). On the other hand, he needs to know whether the bone-growth potential needs the assistance of particular dietary factors; say, vitamin C (for collagen), vitamin D, or extra calcium. It would also, of course, be of great importance to know of any specific agents that stimulate or retard healing, a question to which we shall return.

Nerve Repair as an Example of Healing

Healing, repair, replacement, and regeneration are so closely related that there is no sense in trying to distinguish sharply between them. All are distinguished from maintenance in that they are processes initiated by the occurrence of some incident that is likely to happen but does not recur regularly. The instructions of the organism nevertheless contain provisions to meet such accidents, the occurrence of which is in this sense predicted and expected, although arhythmic. We may use the name as headings under which to describe the repair of damage of increasing severity, by operations ranging from simple clotting of the blood to replacement of a whole limb. Healing has many meanings, but refers especially to such processes as the closure of wounds in the skin or of other organs, where the damge is due not so much to the loss of substance, which may be minimal, as to the interruption of the continuity of the tissues. The effect of severing a nerve provides an interesting example.

The physical break in the nerve trunk damages simply when it interrupts, not because of loss of tissue, but because the contained nerve fibres are dependent for their trophic maintenance on continuity with the nerve-cell body. After severance, therefore, only the central part remains alive. The peripheral axons undergo degeneration and if the two stumps are close together the old axons are then replaced by outgrowth from the central end. We can therefore regard the activities that reunite the stumps as an example of healing and the function of the new fibres as replacement or regeneration. After any wound (in a mammal) the repair can be considered to proceed in three phases: (1) inflammatory, the first-aid operations of closure and removal of debris, (2) proliferative, the advent of fibroblasts, able to lay down new structural materials, and perhaps of specific types of cell; (3) remodelling, such as the maturation of scar tissue or, in nerve repair ,the growth to maturity of some nerve fibres and the atrophy and disappearance of others. The first aid is provided by clotting of the blood. This is produced by a special set of enzymes and other proteins provided for the purpose.

The inactive enzyme prothrombin of the blood is activated by an enzyme thrombokinase, liberated by the damaged tissue, which, in the presence of calcium, converts prothrombin to the active enzyme thrombin. The latter then turns the protein fibrinogen, circulating in the plasma, into threads of fibrin. These help to close the openings of the vessels, stopping the loss of blood. Contraction of the vessels themselves assist this. The fibrin clot serves also immediately to meet the first need of repair, which is to fill up the space left by the wound. But the clot is not itself strong enough to resist any great compression or tension and must be replaced by stronger tissues.

Almost at once, therefore, it becomes infiltrated by white blood corpuscles. The first are small microphages (polymorphs), able to remove debrisi and infective agents; later, larger macrophages move in. together these cells remove the fibrin clot, probably by fibrinolytic enzymes, made available by breakdown of the neutrophil white blood corpuscles and release of their lysosomes. These enzymes work differentially on the fibrin molecules, attacking first those that are not under tension. The clot thus comes to present orientated strands. Along these migrate the fibroblasts, which then proceed to produce collagen to unite the severed surfaces more firmly. When a nerve trunk has been cleanly severed, without other damage, the process of healing varies according to whether the two severe ends remain close together. Let us assume that they do so, or that, having been separated, they have been sewn together by a surgeon. Such clean interruptions may occur when a nerve is cut by glass, say in the wrist when a hand is pushed through a window.

There may often be complications from damage to tendons, and nerves may be equally or more seriously damaged by being pulled, as when a humerus is broken and the radial nerve is bruised by contusion. However, if the severed ends of the nerve are together, a blood clot forms between them. After three or four days the fibrin will be dissolving away and fibroblasts will move in from both the central and distal stumps. Peripheral nerve normally contains much connective tissue, and the fibroblasts presumably come from this, though what activates them will have to be discussed below. The fibroblasts, moving along the strands of fibrin, soon begin to lay down their characteristic protein, collagen, and mucopolysaccharides. If the two joined ends of the nerve are under light tension, then the new collagen fibres will run longitudinally between the ends, in the direction that is most calculated to join them firmly. It is probably desirable, therefore, hat nerve stumps should be left under slight tension, rather than pressed together, which would produce a scar of collagen fibres not all orientated along the line of the nerve.

A similar argument applies even more strongly to the union of the severed ends of a tendon. The strands of fibrin and new strands of collagen serve a further function besides this resistance to stretch. They also guide the course of the outgrowing nerve fibres, or rather of the satellite Schwann cells, which precede the latter. Vertebrate nerve fibres are surrounded by these Schwann cells, which in some make an insulating myelin sheath that greatly increases the speed of nervous conduction. Even those peripheral nerve fibres that have no myelin all have some from of Schwann cell. In regeneration these satellites have a special function. They cease to be ensheathing cells and move out into the site of the wound by active amoeboid movement. They creep along the threads of fibrin and collagen, froming a web of nucleated tissue between the two stumps. They emerge mainly (perhaps wholly) from the lower (distal) stump, where they are released from their enshething function by the gradual breakdown and dissolution of the axons, which have been separated from their cell bodies and are therefore degenerating. A piece of protoplasm separated from its nucleus sooner or later dies. Nerve fibres, being very elongated, are especially vulnerable to this hazard.

The nuclei of the nerve fibres that run to the toes are in the spinal cord and must therefore sustain the protoplasm of threads about a metre long, but only to μm in diameter. It is not known what influence passes along to do this, but there is evidence of an actual transport of material down the fibre. Therefore, 2-3 days after a nerve fibre in a mammal has been interrupted, the isolated part breaks up, first at a few widely separated points and then into a string of avoids and ultimately spheres. The process is not complete for many days; indeed, some pieces of axon remain for several weeks (or longer in cold-blooded animal). Meanwhile the Schwann cells of the peripheral stump divide mitotically, forming strands within the tubes of the supporting tissues of the distal stump. The strands emerge from the cut end and grow up to join the central stump. Here they may make contact with the severed central ends of the nerve fibres. These have formed swollen amoeboid tips, putting out 'feelers' into the tissues between the stumps. When one such feeler touches a strand of Schwann cells it makes some specific from of union and the tip moves away down into the distal stump, spinning a fine thread, the new axon, behind it. We shall discuss its further course later on as an example of regeneration.

Nerve-Growth Factor

Some information is available about chemical factors that stimulate the growth of nerve fibres in the embryo, and these may be related to those of the adult. If tumour cells of mice are implanted in place of the limb bud of a 3-day-old chick the nerve fibres of the latter wander around among the cancer cells. The sensory or sympathetic cells from which the fibres originate are stimulated to increase in number several times so that their nerve fibres invade all the tissues of the host embryo. They even enter the blood-vessels in such numbers as to interrupt the circulation (Levi-Montalcini 1964, Levi-Montalcini and Angeletti 1968). In the search for the chemical nature of the 'nerve growth factor' (NGF) responsible for the increase it was found that snake venom has a great power to produce the phenomenon, and that an even more potent source of NGF is the salivary glands of a male mouse. The substances are both proteins but no identical. That from snake venom has a molecular weight of about 40,000.

The effect of the factor isolated from the male mouse salivary glands and injected into young mice is to produce a specific hypertrophy only of the sympathetic system. All the rest of the nervous system remains unaffected, even the parasympathetic ganglia. A further development was the preparation of an antiserum that *suppressed* growth of the sympathetic cells. The purified NGF protein was injected into other foot-pad of a rabbit. Serum from the rabbit injected into new-born mice then caused the latter to develop with as little as 3 per cent of their normal quota of sympathetic cells, the growth of the rest being suppressed by the antibodies. We cannot yet relate these findings to the processes of normal growth and development of nerve, but they provide further valuable evidence of the existence of specific chemical factors able to stimulate or suppress the growth of particular types of nervous tissue. With further knowledge of such factors it should eventually be possible to understand how the nervous system becomes provided with the many types of channel that constitute the essential feature of its system of coding and signalling.

Processes of Healing

The union of two nerve stumps provides a typical example of the basic processes of wound healing. We might list the following as the components.

1. Quickly filling the defect, achieved by the fibrin clot—as it were, a provisional emergency repair.
2. Next, filling the defect with cells, first polymorphs, then macrophages, fibroblasts, and Schwann cells, and finally the characteristic cells of the tissue, nerve fibres.
3. Giving physical strength to the new tissue by collagen, making good the mechanical damage.
4. Repairing also the vascular defect, by ingrowth of new vessels.
5. Finally, repairing the actual functional defect, in this case by maturation of the nerve fibres to make effective lines of nerve conduction.

It is obvious that many specific growth processes are concerned. A few of them are simple enough for us to be able to see the reactions in some detail. Thus, the clotting of the fibrinogen and formation of the new collagen are the results of change or synthesis of relatively simple proteins. It is much harder to understand how the orderly sequence of so many processes is achieved. In particular, we should like to know what activates the cells to divide and to migrate after wounding and what stops them when their aim has been achieved.

Theories of Wound Healing

The various theories of the stimulus to wound healing offer different levels of explanation.

1. The simplest is that there are wound hormones produced by the damage to the cells. Thee are two versions of this, (a) that the substances stimulate the cells to divide and migrate, (b) that they remove an inhibition on these activities.

2. A second group of theories suggests that as the repair process serves to fill up the defect left by the injury, it is this deficiency or lack that somehow sets in motion the actions of repair.
3. The third theory is that it is the functional lack that is sensed by a suitable system and thus initiates repair.

The second and third approaches seem at first sight much vaguer than the first, but may have more value than appears. In particular, we have to study not only what initiates repair but also what ends it at the correct point. The fact that the defect has been filled or the function restored may provide the relevant information that stops the tissue growing.

It is unfortunately true that at present we do not know what initiates or controls repair mechanisms. This is indeed a great weakness, but biological and medical science are faced here with a fundamental difficulty of method. One cannot distinguish between theories of the three types outlined above because one can not produce a loss of tissue without producing traumatized cells. Conversely, one can not do damage without causing some from of deficit. In plants it has been established that injured cells release a substance, traumatic acid ($HOOC(CH_2)_8CH{=}COOH$), which stimulates cell division near by. No similar relatively simple wound hormones have been identified in animals.

Indeed, the whole question of growth-promoting substances is highly complicated. From the earliest days of tissue culture it was found that growth is stimulated by extracts, especially by extracts of embryos, but also of other tissues. But such extracts are very complex mixtures and it is still not clear to what extent their effects are due to the nutrients they supply or to specific growth stimulants. There is, however, a considerable body of scattered evidence that the growth of each tissue can be stimulated or inhibited by specific substances produced within that tissue. Examples of these substances are the epidermal growth factor in the skin, nerve growth factor, and erythropoietin, which stimulates the bone marrow. Extracts of various adult and embryonic tissues have been shown to stimulate growth in tissue culture under carefully controlled conditions. The stimulating power is associated with protein.

Wound-Healing in the Skin

Repair of damage to the surface has perhaps a special human interest since we all observe it so often in ourselfes. Moreover, its superficial position makes the skin especially easy to study. After damage to the epidermis and underlying dermis there is first the formation of a fibrin clot and then its dissolution after a few days. Cells of the neighbouring epidermis are stimulated to division. They are normally, of course, in a cycle of continuous division, and presumably the mechanism for ensuring the increase is related to this. After 12 to 24 hours the cells move in such a way as to cover the damaged area. Meanwhile the cells of the underlying dermis also proliferate, and polymorphs, macrophages, and fibroblasts proceed to operate as in the nerve regeneration already mentioned. The response of the epidermis is especially interesting. In a small wound its cells continue to proliferate and to migrate until the form a complete covering.

The migration is apparently triggered by the absence of contact between the cells and stops when the space is filled by contact inhibition which stops the fibroblasts. The cells then proceed to contract, drawing the separated edges of the wound together. This process of wound contraction is distinguished from the 'wound contracture' by which collagen fibres may draw together the edges, at a much later stage, in a large, poorly healing wound leaving a bad scar. The

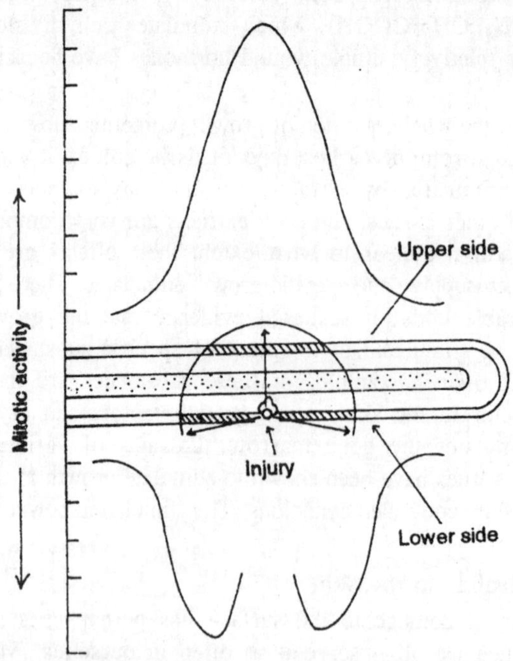

Fig. 6.2. The relation of epidermal hyperplasia to a wound.

mechanism of skin-healing thus shows all the factors we have seen before: filling the space, adding new cells, and restoration of function. Here, however, there is some information about the stimulating factors. The stimulus to cell division is specific to the cells that are damaged. If a flap of mouse skin is cut beneath the dermis, then the overlying epidermis shows no increase in mitosis. Conversely, damage to the epidermis causes mitosis only there and not in the dermis. There is some evidence that the stimulation is produced by an extractable *epidermal growth factor*.

Extrats of epidermis have been reported to increase the rate of epidermal wound healing when applied either locally or by in traperitoneal injection. Particularly convincing evidence has been obtained in the bug *Rhodnius*. A small skin wound is surrounded by a zone of mitotic cells and this area is increased by the application of an autolysed epidermal cell extract. However, the effect is not specific and is produced by a variety of degarded protein extracts. It may be due simply to damage of the cells of the recipient by the extract. So we have again the same problem that the healing response appears only when tissues have been removed. A somewhat similar phenomenon is found in mouse skin. After an epidermal wound the hair follicles at some distance away show mitotic stimulation. This again suggests a chemical effect, but brings up the question of whether this is the production of a stimulant ('promotor') or the depletion of an inhibitor ('depressor') due to loss of tissue.

The theory that removal of depressors is involved has many attractions, since it explains the otherwise puzzling phenomenon that the tissues respond to a *loss* in their neighbourhood, and also suggests the negative feedback by which growth is stopped as the defect is repaired. Evidence that the stimulus is in fact removal of an inhibition has been obtained by Bullugh and Laurence (1960). They found that mitosis extends about 1 mm around a small wound and will occur even on the opposite side of the thin ear of a mouse. But the response of, say, the upper side of the ear occurs only if the epidermis of the under side is *completely* removed. It is not, therefore, a reaction to damaged epidermal cells but to removal of some influence previously inhibiting mitois. The concept is that the; cells are prevented from mitosis by the inhibitory action of substances they themselves produce ('chalones').

Further support has come from the study of wounds of the lens epithelium of the rabbit. Twelve hours after wounding, cells in the neighbourhood of the injury begin to synthesize DNA (as shown with tritiated thymidine). Shortly afterwards they divide. The region of DNA activation and division then spreads peripherally away from the wound. The explanation seems to be that after injury cells migrate into the wound. This depletes the neighbouring zone, reducing the concentration of inhibitor and setting in motion the procedure of DNA replication. Meanwhile cells further out have moved in, leading to a wave of depletion and hence to a wave of mitosis that is propagated outwards at about 17 μm/h. Twenty-four hours later the cells near the wound again incorporate DNA and a second wave passes outwards. This is indeed a possible interpretation, but there is no direct evidence of a chemical inhibitor.

Another possibility is that the cells begin to synthesize DNA only when they have lost contact with their neighbours, as happens in some tissue cultures. We thus have the outlines of a hypothesis about the control of repair in epidermal structures. Once again, the replication of DNA is seen to be the central phenomenon. The process of repair appears to be triggered by the most obvious stimulus, namely the absence of a component previously present in the neighbourhood. If this is due to the absence of a substance normally produced by the cells themselves, then we have a mechanism by which the repair is terminated when the appropriate concentration has been restored. It is unfortunate, however, that the specific substances concerned have not been isolated or identified. But there is enough evidence to make it well worth while to continue the search for suitable sources of them. We now have to inquire whether similar principles apply to replacement of larger quantities of material and to regeneration of whole organs.

Replacement and Regeneration of Parts after Loss

Regeneration of the Liver and Other Glands

The replacement of a single mass of a tissue such as liver is easy to study experimentally, although perhaps a not very common event in nature, except through disease. In tissue that renew by cell multiplication (such as the epidermis) the increased production needed for regeneration is presumably started off by acceleration of whatever system controls daily replacement. In the so-called 'expanding' tissues, however, such as the liver, where mitotic activity in the adult is normally very low, special mechanisms must be brought into play to produce replacement after loss of a large mass. These mechanisms are indeed exceedingly powerful. After reduction of the liver of a mouse to as little as 30 per cent of its mass it will return to normal within 6-8 days. A kidney will enlarge if its fellow is removed, and a host of similar compensatory responses can occur throughout the body. We may perhaps distinguish situations in which the loss affects a systemic function, and produces hypertrophy elsewhere inthe body, from purely local losses.

Thus, after removal of the salivary glands of a rat on one side those of the other side hypertrophy. Such examples are numerous and suggest that the response to removal is basically governed by functional consideratins. This, however, does not give us an insight into the mechanisms concerned, except to suggest that they are specific to each tissue and have been highly selected. Indeed, action of these compensatory processes. The liver of rats has been especially well investigated. After partial hepatectomy the cells and nuclei enlarge and at about 18 hours after removal mitosis begins. It quickly reaches a maximum of nearly 3 per cent of the cells (normal 0.01 per cent) and then after about 30 hours begins to decline towards normal. The original weight is recovered withina week.

The functions of the liver are so varied that it has been impossible to decide which of them it is whose deficiency triggers the cell multiplication. It is probably partly the lowering of the plasma protein and the induction of specific protein synthesis. Other factors may be the failure of the digestive functions, or simply a result of the distortion of the; pattern of vascular flow. Similarly, after kidney removal there are many physiological repercussions, any of which may trigger the hypertrophy of the remaining kidney. This is not due simply to the need for the single kidney to produce more urine. One ureter can be diverted to enter the peritoneal cavity or the gut. Its secretion is absorbed and makes more work for the other kidney, but the latter does nto hypertrophy.

Simlar unexplained discrepancies are abundant inthe literature of replacement. Thus one effective way of producing hypertrophy of the salivary gland of the rat is repeated amputation of the lower incisor and the effect is unilateral on the salivary gland only of the same side. Here some very special mechanisms (presumably nervous) must be involved to relate the secretion to the wear on the tooth. The wear would usually be symmetrical and it is an acident that the mechanism has remained unilateral.

Compensatory growth of the endocrine organs is very marked and is normally produced by stimulation from other endocrines. Thus the pituitary stimulates the thyroid, adrenal cortex, sex glands, and others.

In each case any deficiency produced by removal s timulates the stimulator, but the details of how it does so are obscure. Increased activty alone does not necessarily produce increase, as is shown by stimulating an exocrine gland repeatedly, say wtih a drug. Neither the pancreas nor the salivary glands hypertrophy after repeated stimulation by pilocarpine. Curiously enough, another drug that stimulates salivary secretion, isoproterenol, does produce hypertrophy, so that the glands may become five times larger than normal. These examples really serve only to emphasize our ignorance of the mechanism that sets off the hypertrophy.

Functional Regeneration in the Nervous System

Special processes are available for repair of some parts of the nervous system in all animals and of major parts in simpler animals. This is the extreme example of repair of a tissue with fixed cell-number. In mammals, defects of the peripheral nervous system are made good by outgrowth from the existing neurons, without any production of new ones. It should be possible, therefore, to see the factors that are at work initiating and controlling the 'hypertrophy', uncomplicated by cell division. It is indeed true that we can discern the principles of control but there is no detailed knowledge of their chemical basis. Although nervous tissue is post-mitotic it probably undergoes continual replacement even in the absence of injury. As in other tissues, therefore, the regenerative processes are probably an extension of those normally at work.

The axons of the central stump swell to make irregular masses of material, probably of amoeboid nature, from which filaments are spun along the; bands of Schwann cells to reach the peripheral stump. Once within the tubes of the stump they advance at a rate of 5 mm/day (in mammals) and are robably led by the tube all the way to a peripheral organ. However, the question of how correct connections are made is still highly controversial. The nervous system differs from all others in the extent of its division into specifically different units, which must be correctly connected to function properly. Effective regeneration therefore depends upon the outgrowing fibres making connection with the right peripheral tubes. Ths is assuming that there is no other means of finding the right peripheral organ or alternatively of remodelling the central connections. The latter is improbable, at least in adults of higher organisms. On the other hand, there is strong evidence of special mechanisms for ensuring correct connections at the end of the axon.

The optic nerve can regenerate inmany vertebrates, from fishes to birds. The new nerve fibres may become much interwoven in the nerve trunk, but each of them ultimately comes to re-make connections with its original partners inthe optic lobe of the mid-brain. Unfortunately there is really no satisfactory view as to how this astonishing and very important result is produced, and it would nto be profitable to discuss the interesting evidence here. It emphasizes that very precise mechanisms must be at work. A little is known about the types of factor that operate for the control of the development of new fibres in a regenerating peripheral nerve stump. In a normal nerve the fibres with different functions differ in diameter. Thus the fibres that carry signals to the muscles to make them contract are large, and conduct fast.

Afferent fibres that carry signals about the tension on the muscle are also large. Fibres signalling touch are rather smaller. Fibres controlling blood flow (sympathetic) are very small and conduct slowly, presumably because they change the state of the blood-vesels only slowly and maintain them in a given s tate for relatively long periods. When new nerve fibres are first laid downtheyare all very thin, whether in the embryo or in a regenerating nerve. Some remain small, others grow larage and acquire sheaths, the process of 'maturation'. There must be specific factors at work to control the growth of each to the correct size. In fact each type of nerve fibre is highly specified from some early embryonic stage, probably by virtue of the connections made at the periphery, though we know little indetail abotu this.

During regeneration the process is at least to some extent repeated. For example, motor firbes that do not reach to their normal periphery do not return to their original size. Some influence must pass *up* the nerve from the muscle or other end organ to control the synthesis of new axoplasm. Unfortunately, once again we must admit to complete ignorance of the nature of this influence. New axoplasm is probably produced mainly in the nerve-cell body. After the axon has been interrupted there are quite explosive changes in the RNA and synthetic mechanism in the cell, as indicated by the process known as retrograde degeneration or chromatolysis. This again shows that some reverse signalling takes place, by which the damaged lower end of the neuron evokes the production of extra supplies of axoplasm.

Presumably a further set of signals switches off the synthetic proceses when sufficient axoplasm has been produced to return the fibre to its appropriate diameter, large or small. The end result of the nerve regeneration is thus to fill the space left as a result of the injury. However, the effectivensss of the repair lies in the capacity for useful signalling that it confers. Whether as pecific mechanism will be present to repair any part of the nervous system therefore depends on whether repair is likely to increase the survival value of the individual and of the species. Regeneration of the peripheral nervous system is no doubt party associated with a normal wear and tear process, for example at the tips of the nerves in the skin of the fingers.

Interruption of a few small peripheral nerve fibres will not be sufficiently in capacitating to make life impossible, and repair processes to make the part useful again are well worth while. On the other hand, damage to the central nervous system is probably such a serious handicap that it is not worth while to attempt to make it good. A mammal or bird with the spinal cord severed is so severely handicapped that in nature it could never stay alive long enough for regeneration to take place. Indeed, it is surprising that it is worthwhile for many fishes to have the capacity to regenerate the CNS when the spinal cord is severed. Perhaps it is because the swim-bladder and the waterprovide support that is lackig on land. It is interesting that frogs also lack power to regenerate the spinal cord whereas this is present in newts and tadpoles. Similarly, the higher nervous centres are so important to the life of a bird or mammal that survival is impossible without them and there is no mechanism for their repair. That the fore-brain and rectum can be regenerated in some fishes and newts thus really shows how little use the animal makes of them, since it can survive for a while without them! It may be asked what form of damage to the brain is likely to leave a fish in a condition in which repair is worth while. Presumably not after mechanical injury, for if the jaws of a redator damaged the brain, they would hopelessly disorganize the skull. It may be that the power of regeneration in these animals has developed to repair the damage of parasites rather than predators.

The brains of fishes are often found to be packed with helminth (worm) parasites. The brains of higher vertebrates apparently have methods that prevent the entry of such invaders and it may be for this reason that they have lost the power of regeneration. But of course with increasing complexity and detail of connections the morphogenetic processes needed for regeneration must become exceedingly complex. The capacity to regenerate the optic nerve in many vertebrates, up to and including birds, shows presumably that one eye can compensate for the missing one for a time, but also that to have both significantly increases the probability of survival. To understand this situation properly we should need to know more of how the animals use their eyes in the normal environment.

Regeneration of Limbs

The same principle applies to regeneration of limbs and of internal organs. Mechanisms able to produce repair have been evolved for organs that are useful but not indispensable for the life of the individuals. This raises interesting questions about the selective value of carrying whatever load is imposed by processes that are used only intermittently. Of course the rate of loss varies greatly. Ninety-three per cent of individuals of some species of crustaceans have been found to show loss of major parts. Siince crabs and lobsters have many liimbs they can afford to be without one while it is growing. There is little dubt, from the fact that the power of regeneration is used so often, that is has survival value. It is sometimes suggested that this capacity is merely a 'casually persisting pristine power of morphogenesis.' It is much mor elikely that the basic powers of metabolic turnover have been specially selected and improved upon to produce regeneration of the more vulnerabel members. Proof that these are specially selected powers comes from the animals such as lizards that have evolved autotomy mechanisms, regions at which a limb or tail is cast off when it is seized and remains wriggling to distract the attacker. There are special mechanisms to close the wound, and regeneration takes place faster following autotomy than after injury elsewhere. Autotomy may be useful even when not followed by regeneration. The crane-flies ('daddy-longlegs', tipulids) often cast off a leg when captured, but cannot replace it.

Regeneration and Evolutionary Advance

It is widely held that lower organisms have greater regenerative powers than higher ones. The attempt to produce quantitative scales either for evolutionary status or regenerative power is hazardous but has been bravely made by Needham (1964). He concludes that regenerative power is inversely related to evolutionary grade and that the reason is probably that morphological, and particularly histological, differentiation are directly related to the grade, so that if de-differentiation is necessary as a prelude to regeneration it becomes increasingly difficult or uneconomic. He recognizes, however, that the matter is very complicated and many factors may influence the presence or absence of regenerative power. Thus the alertness and nimbleness of the higher animals as potential prey saves them from serious wounds short of capture and total destruction, for which the appropriate biological defence is not regeneration but reproduction. On the other side there are indications that the regenerative power is not *necessarily* lost in all higher animals.

The capacity of stags to regenerate huge antlers each year is a conspicuous example. It seems unreasonable that the skin of a small area should have the power to produce the luxuriant velvet of new hair follicles, much more perfect tan the skin that is regenerated over wounds. Yet this beautiful new covering is then 'rubbed off in tatters soon after antler growth is complete'. It is clear that we are still a long way from understanding the biological significance of many phenomena of growth and regeneration. We think that we can appreciate repair processes that are of value to ourselves, such as healing of wounds or of bones. But we know so little of the significant features of the life system of other species that some of their major repairs seem to us to be simply a waste. It is doubtful if this is a sound judgement.

7

POLYGENIC MECHANISM

The Mendelian genetics of beak size in finches is unknown, but the character shows evolutionary changes as the food supply changes through time. We begin by looking at beak size in Darwin's finches as an example of the kind of character studied by quantitative genetics. We then move on to the theoretical apparatus used to analyze characters that are controlled by large numbers of unidentified genes. The influences on these characters are divided into environmental and genetic factors, and the genetic influences are divided into those that are inherited and influence the form of the offspring and those that are not. A number called heritability expresses the extent to which parental attributes are inherited by offspring. With the theoretical apparatus in place, we can then apply it to a number of evolutionary questions, including directional selection, in both artificial and natural examples, and stabilizing selection. We investigate whether a balance between selection and mutation can explain the amount of genetic variation seen in nature, and consider an experiment in which the effect of selection is minimized to reveal the power of deleterious mutation.

Fourteen species of Darwin's finches live in the Galapagos archipelago, and many of them differ most obviously in the sizes and shapes of their beaks. A finch's beak shape, in turn, influences how efficiently it can feed on different types of food. Grant has been studying these finches since 1973, and his best evidence to demonstrate that beak size influences feeding efficiency comes from a comparison of two species, the large-beaked *Geospiza magnirostris* and the smaller *G. fortis*, feeding on the same kind of hard fruit.

The large-beaked *G. magnirostris* can crack the fruit (called the *mericarp*) of caltrop (*Tribulus cistoides*) transversely, taking on average only 2 seconds and exerting an average force of 26 kgf. It can then easily eat all 4-6 seeds of the smashed fruit, taking only 7 seconds. The smaller *G. fortis* is not strong enough to crack *Tribulus* mericarps and instead twists open the lower surface, applying a force of only 6 kgf and taking 7 seconds on average to reach the seeds inside. Only one or two of the seeds can be obtained in this way, and it takes an average of 15 seconds to extract them. Thus, *Geospiza magnirostris* usually has an advantage with these large, hard types of food.

Smaller finches are probably more efficient with smaller types of food, although this capability is more difficult to show. Both large and small finches on the Galapagos eat small seeds, although there is an indirect reason to believe that smaller finches do so more efficiently. From the evidence we have examined so far, we should predict that natural selection would favour larger finches when large fruits and seeds are abundant. The prediction should apply both within and between species. A *G. magnirostris* looks like an enlarged *G. fortis*, and a larger individual *G. fortis* can probably deal with a large food item more efficiently than can a smaller conspecific, much as an average specimen of *G. magnirostris* is more efficient than an average *G. fortis*. When large seeds are common, we might expect the average size of a population of *G. fortis* to increase between generations, and vice versa when large seeds are rare—if beak size is inherited.

If beak size is inherited . . . but is it? Beak size is inherited if parental finches with larger than average beaks produce offspring with larger than average beaks. Grant measured the sizes of parental and offspring finches in several families and plotted the latter sizes against the former. Large-beaked parental finches do, indeed, produce large-beaked offspring: thus, beak size is inherited. It therefore makes sense to test the prediction that changes in beak size should follow changes in the size distribution of food items. The test was carried out by Grant for the species *Geospiza fortis*, on one of the Galapagos Islands, Daphne Major. Since the study began, this species has undergone two major, but contrasting, evolutionary events.

In the Galapagos, the normal pattern of seasons calls for a hot, wet season from about January to May to be followed by a cooler, dryer season through the rest of the year. In early 1977, for some reason, the rain did not fall. Instead of the normal progression of a wet season followed by a dry season, then another wet season and dry season, the dry season that began in mid-1976 continued until early 1978—one whole wet season was missed. The finch population of Daphne

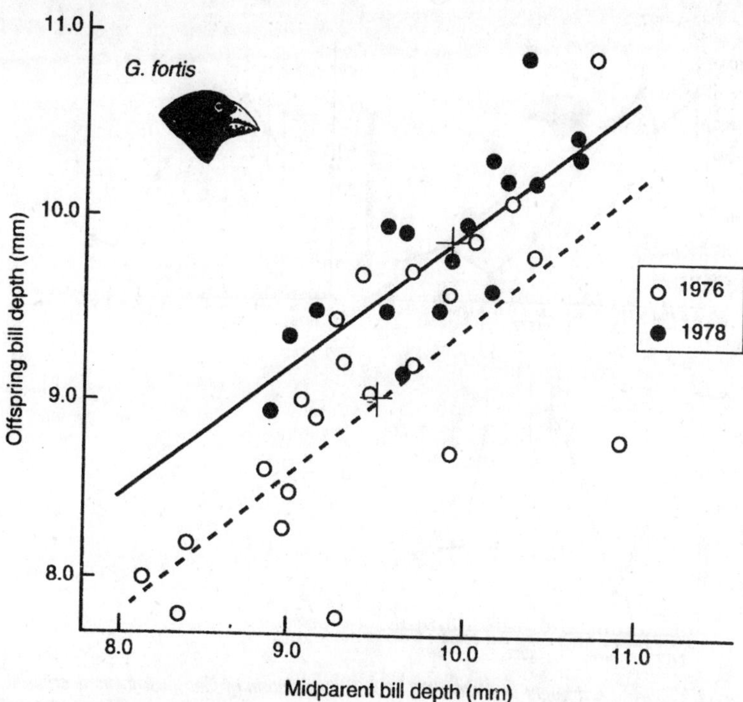

Fig. 7.1. Parents with larger than average beaks produce offspring with larger than average beaks in Geospiza fortis on Daphne Major.

Major collapsed from about 1200 to about 180 individuals, with females being particularly hard hit; at the end of 1977, the sex ratio was about five males per female. As the sex difference shows, not all finches suffered equally. Smaller birds died at a higher rate. The reason, again, relates to the food supply. At the beginning of the drought, the various types of seeds were present in their normal proportions. *G. fortis* of all sizes take small seeds, and, as the drought persisted, these smaller seeds were relatively reduced in numbers. With time, average available seed size became larger. Now the larger finches were favoured, because they could eat the larger, harder seeds more efficiently; and the average finch size increased as the smaller birds died off. (Females died at a higher rate than males because females are on average smaller.) Size, as we have seen, is inherited. The differential mortality in the drought, therefore, caused an increase in the average size of finches born in the next generation. *G. fortis* born in 1978 were about 4% larger on average than those born before the drought.

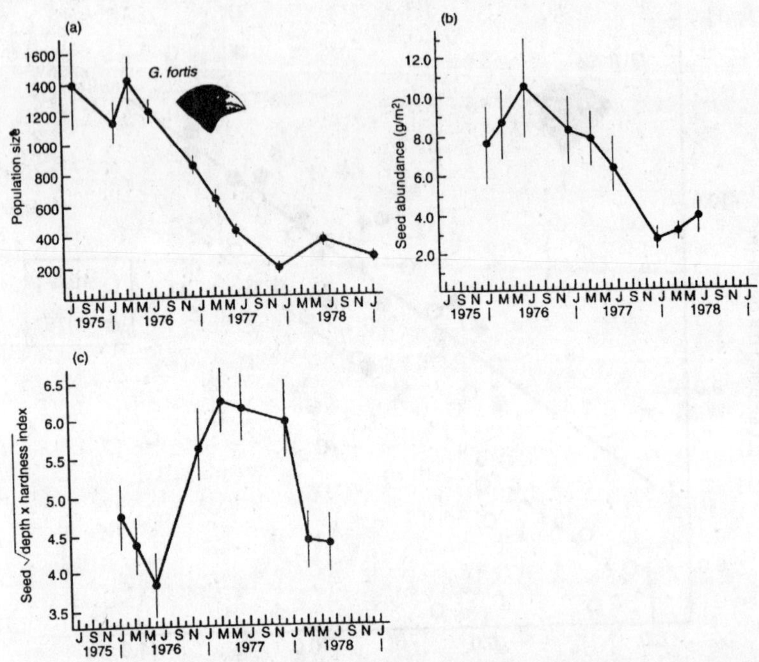

Fig. 7.2. During a drought in 1976-1977, (a) the population of Geospiza fortis decreased on the island of Daphne Major in the Galapagos archipelago, due to (b) the decline of the food supply. (c) The average size of the seeds available as food increased during the drought.

Four years later, in November 1982, the weather reversed. The rainfall of 1983 was exceptionally heavy and the dry volcanic landscape became covered with foliage in the periodic disturbance called El Nino. Seed production was enormous. The theory developed for 1976-1978 could now be tested. The conditions had reversed: the direction of evolution should go into reverse as well. In the year after the 1983 El Nino event, an excess of small seeds appeared. If the smaller finches could, in fact) exploit them more efficiently, the smaller finches should survive relatively better. Grant again measured the sizes of *G. fortis* on Daphne Major in 1984-1985 and found that the smaller birds were indeed favoured; finches born in 1985 had beaks about 2.5% smaller than those born before the El Nino downpours. His theory, that seed sizes control beak size in these finches, was confirmed.

The fluctuations in the direction of selection on beak shape—with smaller beaks favoured in some years and larger ones in others—probably results in a kind of "stabilizing" selection over a long period of time such that the average size of beak in the population is the size favoured by long-term average weather. Later in the chapter, we will see how the degree of selection can be expressed more exactly.

Quantitative Genetics

The beak size of Galapagos finches illustrates a large class of characters that shows *continuous variation*. Simple Mendelian characters, like blood groups or the mimetic variation of *Papilio*, often have discrete variation. Many of the characters of species are like beak size in these finches, varying continuously, such that every individual in the population differs slightly from every other individual. There are no discrete categories of beak size in *G. fortis* or in most other species of birds.

The other important point about beak size is that we do not know the exact genotype that produces any given beak size. We can, however, say something about the general sort of genetic control it may have. Characters like beak size, which has an approximately normal frequency distribution (i.e., a bell curve), are probably controlled by a large number of genes, each of small effect. Imagine that beak size was controlled by a single pair of Mendelian alleles at one locus, with one dominant to the other, so that *AA* and *Aa* is long and *aa* is short. In this case, the population would contain two categories of individuals. Now suppose that beak size was controlled by two loci with two

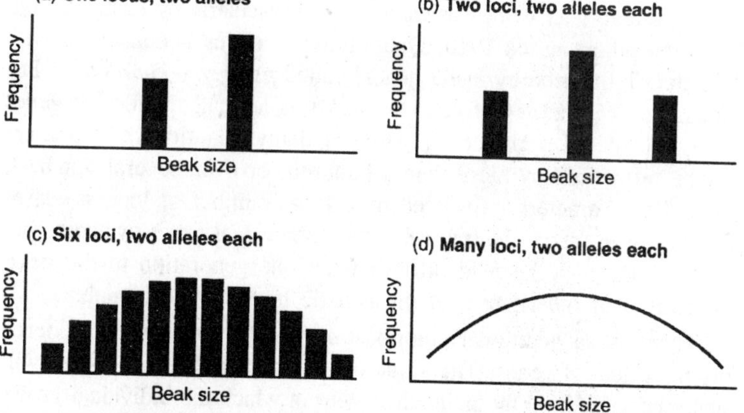

Fig. 7.3. (a) The phenotypic character—beak size, for example—is controlled by one locus with two alleles (A and a). A is dominant to a. (b) The character is controlled by two loci with two alleles each; A and B are dominant to a and b. (c) Control by six loci, with two alleles each. (d) Control by many loci, with two alleles each.

alleles each. Beak size might now have a background value (e.g., 1 cm) plus the contribution of the two loci; an *A* adds 0.1 cm and a *B* subtracts 0.1 cm. *A* and *B* are dominant to *a* and *b*. Then *aaBb* and *aaBB* individuals have 0.9 cm beaks, *aabb*, *AaBB*, *AaBb*, *AABB*, and *AABb* have 1 cm beaks, and *AAbb* and *Aabb* have 1.1 cm beaks. The frequency distribution if all alleles have frequency one-half and the two loci are in linkage equilibrium. The distribution now has three categories and has become more spread out. It becomes even more spread out if beak size is influenced by six loci and it begins to look normal for 12 loci.

Thus, when a large enough number of genes influence a character, it will have a continuous, normal frequency distribution. The normal distribution can result if a large number of alleles are present at each of a small number of loci influencing the characters, or if fewer alleles appear at a larger number of loci. In this chapter, we will mainly discuss the theory of quantitative genetics as if there were many loci, each with a small number of alleles. Although this genetic system may well underlie many continuously varying characters, the theory applies equally well with a few (even one) loci containing many alleles at each.

Mendel had noticed in his original paper in 1865 that multifactorial inheritance (i.e., the character is influenced by many genes) can generate a continuous frequency distribution. This theory was not well confirmed until later work was carried out, particularly by East, Nilsson-Ehle, and others, circa 1910. Quantitative genetics is concerned with characters influenced by many genes, called *polygenic characters*. For a quantitative geneticist, five to 20 genes is a small number of genes to be influencing a character; whereas many quantitative characters may be influenced by more than a hundred, or even several hundred, genes. For characters influenced by a large number of loci, it ceases to be useful to follow the transmission of individual genes or haplotypes (even if they have been identified) from one generation to the next. The pattern of inheritance, at the genetic level, is too complex.

There is an additional complication. So far we have considered only the effect of genes. The value of a character, like beak size, will usually be influenced by the environment in which the individual grows up. Beak size is probably related to general body size, and all characters related to bodily stature will be influenced by the amount of food an organism happens to find during its life. If we take a set of organisms with identical genotypes and allow some to grow up with abundant

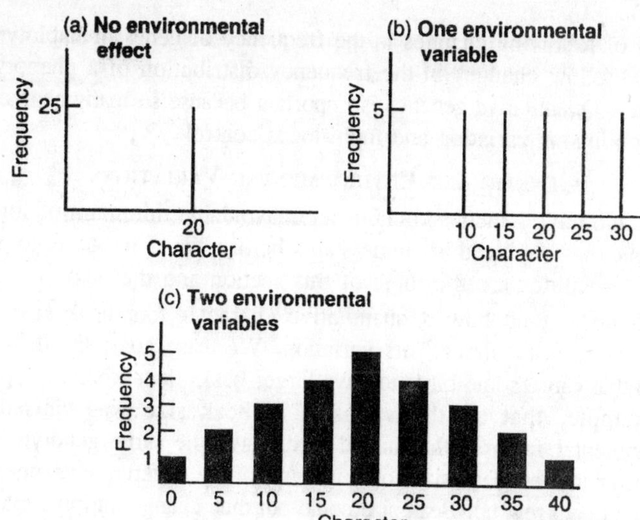

Fig. 7.4. Environmental effects can produce continuous variation. (a) Twenty-five individuals in the absence of environmental variation all have the same phenotype, with a value for a character of 20. (b) Influence of one environmental variable. (c) Influence of a second environmental variable.

food and others with limited food, the former will typically become larger. In nature, each character will be influenced by many environmental variables, some tending to increase it, others to decrease it. Thus, if we consider a class of genotypes with the same value of a character before the influence of the environment and add the effect of the environment, some of the individuals of each genotype will be made larger and others smaller in various degrees. This variation will produce a further "spreading out" of the frequency distribution. Any pattern of discrete variation in the genotype frequency distribution is likely to be obscured by environmental effects and the discrete categories converted into a smooth curve.

The small effects of many genes and environmental variables are two separate influences that tend to convert the discrete phenotypic distribution of characters controlled by single genes into continuous distributions. A continuous distribution for a character could, in principle, result from either process; quantitative genetics is mainly concerned with characters influenced by both factors. Quantitative genetics employs higher-level genetic concepts that are genetically less exact than those of one- or two-locus population genetics, but which are more useful for understanding evolution in polygenic characters.

Instead of following changes in the frequency of genes or haplotypes, we now follow changes in the frequency distribution of a phenotypic character. Quantitative genetics is important because so many characters have continuous variation and multi-locus control.

Genetic and Environmental Variations

Quantitative genetics contains an unavoidable minimum of formal concepts that we need to understand before we can put it to use. Those formalities are the topic of this section and the next.

To understand how a quantitative character like beak size will evolve, we must "dissect" its variation. We tease apart the different factors that cause some birds to have larger beaks than others. Suppose, for example, that all the variation in beak size was caused by environmental factors—that is, all birds have the same genotype and they differ in their beak sizes only because of the different environments in which they grew up. Beak size could not then change during evolution (except for non-genetic evolution due to environmental change). For the character to evolve, it must be at least partly genetically controlled. We need to know how much beak size varies for environmental reasons, and how much for genetic reasons. However, even if different finches vary in their beak size for genetic reasons, it does not necessarily mean it can evolve by natural selection. As we will see, genetic influence must be classified even further, into components that allow evolutionary change and those that do not.

In quantitative genetics, the value of a character in an individual is always expressed as a deviation from the population mean. Thus, beak size will have a certain mean value in a population, and we can talk about environmental and genetic influences on an individual as deviations from that mean. The procedure is easy to understand if the population mean is viewed as a "background" value, and the influences leading to a particular individual phenotype are then expressed as increases or decreases from that value. Suppose there is one locus with two alleles influencing beak depth. *AA* and *Aa* individuals' beaks are 1 cm from top to bottom, *aa* individuals 0.5 cm. In this system, the environment has no effect. If the population average was 0.875 cm (as it would be for gene frequency of $a = {}^1/_2$), then the beak phenotype of *AA* and *Aa* individuals would be written as +0.125 cm and that of *aa* as -0.375 cm. In general, we symbolize the phenotype by *P*. In this case, $P = +0.125$ for *AA* and *Aa* individuals and $P = -0.375$ for *aa*.

Clearly, the value of *P* for a genotype depends on the gene frequencies. When the frequency of *A* is ${}^1/_2$, the genotypic effects are those just given. When the frequency of *A* is $^1/_4$, however, the population mean would be 0.71875 cm. For *aa*, *P* is now –0.21875; for *AA* and *Aa*, $P = +0.28125$.

In this example the phenotype is controlled only by a genotype, whose effect can be symbolized by *G*. *G*, like *P*, is expressed as a deviation from the population mean. In this case, for an individual with a particular genotype (because the environment has no effect)

$$\text{mean of population} + P = \text{mean of population} + G$$

The background population mean cancels from the equation, and can be ignored. We are then left (in this case in which the environment has no effect) with $P = G$.

The value of a real character will usually be influenced by the individual's environment as well as its genotype. If the character under study is related in some way to size, for example, it will probably be influenced by how much food the individual found during its development, and how many diseases it has suffered. These environmental effects are measured in the same way as for the genotype, as a deviation from the population mean. If an individual grew up in an environment that caused it to develop a larger beak than average, its environmental effect will be positive. In contrast, the environmental effect would be negative if it grew up in an environment giving it a smaller than average beak. The phenotype can then be expressed as the sum of environmental (*E*) and genotypic influences:

$$P = G + E$$

This simple relationship is the fundamental model of quantitative genetics. For any phenotypic character, the individual's value for that character (expressed as a deviation from the population mean) results from the effect of its genes and environment.

We must look further into the genotypic effect. We need to consider how to subdivide the genotypic effect, and why the subdivision is necessary. The main point can be seen in the one-locus example we have already used. The *A* gene is dominant, and both *AA* and *Aa* birds have 1 cm beaks ($P = +0.125$). (Because we are investigating the genotypic effect, we will ignore environmental effects, so that $P = G$.) Suppose we take an *AA* individual and mate it to another bird drawn at random from the population. The gene frequency = $^1/_2$ and the random bird is *AA* with chance $^1/_4$, *Aa* with chance $^1/_2$, and *aa* with chance $^1/_4$. Regardless of the mate's genotype, all offspring will have beak phenotype $P = +0.125$, because *A* is dominant. Now suppose we take an *Aa* individual and mate it to a random member of the population. The average beak of their offspring is ($^1/_4$ × +0.125) + ($^1/_2$ × 0) + ($^1/_4$ × −0.125) = 0. That is, the average offspring beak size is the same as the population average. Thus, for two genotypes with the same beak size (*P* = +0.125), one produces offspring with beaks like their parents, while the other produces offspring with beaks closer to the population average.

As we can see, some genotypic effects are inherited by the offspring and some are not. We can divide the genotypic effect into a component that is passed on and a component that is not. The component that is passed on is called the *additive effect* (*A*) and the component that is not is called (in this case) the *dominance effect* (*D*). The full genotypic effect in an individual is the sum of the two:

$$G = A + D$$

The additive effect has the greatest impact. The parent deviates from the population mean by a certain amount, and its additive genotypic effect is the part of that deviation potentially passed on to its offspring. However, when an individual reproduces, only half of its genes are inherited by its offspring. The offspring inherit only half the additive effect of each parent. Thus, the additive effect *A* for an individual is equal to twice the amount by which its offspring deviate from the population mean, if mating is random. For the *AA* parent, therefore, the additive effect is +0.25. (The full quantitative genetics of the *AA* individuals are: *G* = +0.125, *A* = +0.25, and *D* = −0.125.) The offspring of *Aa* birds deviate by zero, so their additive effect is twice zero (or simply zero). Thus, the amount by which *Aa* heterozygotes deviate from the population mean is entirely due to dominance, and is not inherited by their offspring. (For *Aa* individuals, *G* = +0.125, *A* = 0, *D* = +0.125.)

The division of the genotypic effect into additive and dominance components tells us what proportion of the parent's deviation from the mean is inherited, and reveals how the non-inheritance of the *Aa* individuals' genotypic effect is due to dominance. In practice, quantitative geneticists do not know the genotypes underlying the characters they study; they know only the phenotype. They might, for instance, focus on the class of birds with 1 cm (*P* = +0.125) beaks. In the example given above, the additive component of their phenotypic value depends on the frequencies of the *AA* and *Aa* genotypes. If all birds with 1 cm beaks are *Aa* heterozygotes, then none of the offspring will inherit their parents' deviation; if they are all *AA*, then half of the offspring will inherit the deviation.

Why is the additive effect of a phenotype so important? Once the additive effect for a character has been estimated, that estimate plays much the same role in quantitative genetics as the exact knowledge of Mendelian genetics in a one- or two-locus case. We use this estimate to predict the frequency distribution of a character in the offspring, given a knowledge of the parents. In a one-locus genetic model, we know the genotypes corresponding to each phenotype, and can predict the phenotypes of offspring from the genotypes of their parents. In the case of selection, the gene frequency in the next generation is easy to predict if we know selection allows only *AA* individuals to breed. In two-locus genetics, the procedure is the same. If the next generation is formed from a certain mixture of *Ab*/*AB* and *AB*/*AB* individuals, we can calculate its haplotype frequencies if we know the exact mixture of parental genotypes.

In quantitative genetics, we do not know the genotypes. We have only measurements of phenotypes, like beak size. If we can estimate the additive genetic component of the phenotype, then we can predict the offspring in a manner analogous to the procedure when the real genetics are known. When we know the genetics, Mendel's laws of inheritance tell us how the parental genes are passed on to the offspring; when we do not know the genetics, the additive effect tells us which component of the parental phenotype is passed on. Thus, estimating the additive effect is the key to understanding the evolution of quantitative characters. The estimates are practically made by breeding experiments. In the case of finches with 1 cm beaks in a population of average beak size 0.875 cm, the additive effect can be measured by mating finches with 1 cm beaks to random members of the population. The additive effect is then two times the offspring's deviation from the population mean.

The genetic partitioning that we have undertaken so far applies reasonably well for one locus. In that case, dominance is the main reason why the genotypic effect of a parent is not exactly inherited in its offspring. When many loci influence a character, *epistatic interactions* between alleles at different loci can produce additional effects. Epistatic interactions, like dominance effects, are not passed on to offspring—they are not additive. They depend on particular combinations of genes and the effect disappears when the combinations are broken up (by genetic recombination). An example of an epistatic interaction would be for individuals with the haplotype A_1B_2 to show a higher deviation from the population mean than the combined average deviations of A_1 and B_2; the extra deviation is epistatic. Other non-additive effects can arise because of gene-environment interaction (when the same gene produces different phenotypes in different environments) and gene-environment correlation (when particular genes are found more often than random in particular environments).

A full analysis can take all of these effects into account. In a full analysis, such as the simple case given here, the aim is to isolate the additive effect of a phenotype. The additive effect is the part of the parental phenotype that is inherited by its offspring.

VARIANCE OF CHARACTERS

We return now to the frequency distribution of a character. We will continue to express effects as deviations from the population mean. If we consider an individual some distance from the mean, some of its deviation will be environmental, some genetic. In the genetic component, some parts will be additive, some dominant, some epistatic. These terms have been defined so that they add up to give the exact deviation of the individual from the mean. Any individual has its particular phenotypic value (P) because of its particular combination of environmental experiences and the dominance, additive, and interaction effects in its genotype. The different combinations of E, D, and A in different individuals explain why the character shows a continuous frequency distribution in the population.

The variation seen for the character in any one population could exist because of variation in any one of, or any combination of, these effects. Thus, the individual differences could all be due to different environmental effects, with every individual having the same value for G. Alternatively, it could be 25% due to the environment, 20% due to additive effects, 30% due to dominance effects, and 25% due to interaction effects. The proportion of variation related to the different

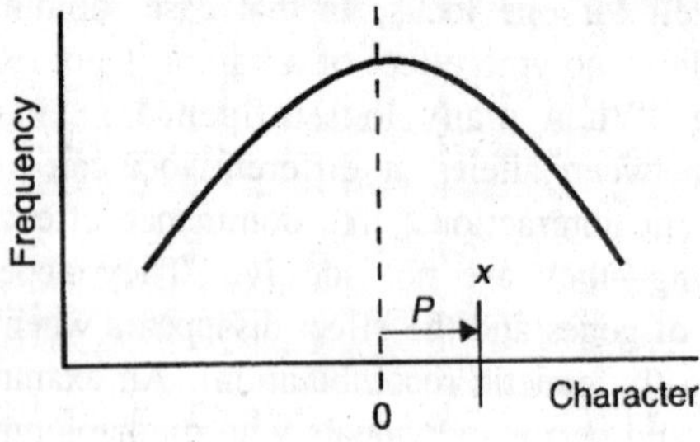

Fig. 7.5. The x-axis of the continuous distribution for a character can be scaled to have a mean of 0.

effects matters when we consider how a population will respond to selection. If the variation exists because different individuals have different values of E, selection will not generate a response. On the other hand, if the variation is mainly additive genetic variation, the response will be large. The proportion of the variation resulting from different values of A in different individuals tells us whether the population can respond to selection.

The variability in the population related to any particular factor, such as the environment, is measured by the statistic called the *variance*. Variance is the sum of squared deviations from the mean divided by the sample size minus one. For all values x of a character like beak size in a population, the variance of x is

$$V_x = \frac{1}{n-1}\Sigma(x_i - \bar{x})^2$$

We have seen how the total phenotype, genetic effect, environmental effect, and so on, can be measured for an individual. The measurements (P for phenotypic effect, and so on) are expressed as deviations from the mean. We can easily calculate what their variances are for a population.

phenotypic variance $= V_P = \frac{1}{n-1}\Sigma P^2$

environmental variance $= V_E = \frac{1}{n-1}\Sigma E^2$

genetic variance $= V_G = \frac{1}{n-1}\Sigma G^2$

dominance variance $= V_D = \frac{1}{n-1}\Sigma D^2$

additive variance $= V_A = \frac{1}{n-1}\Sigma A^2$

The phenotypic variance for a population, for example, expresses how spread out the frequency distribution for the character is. If it is spread out, with different individuals having very different values of the character, the phenotypic variance will be high. If the distribution is a narrow spike, with most individuals having a similar value for the character, phenotypic variance will be low.

Sibling Genotypes

Grant measured the depth of the beak in many parent-offspring pairs in *Geospiza fortis*, and plotted the results on a graph. Each point indicates one offspring and the average of its parents. Beak depth in *G. fortis*, like many characters in most species, shows a correlation between parent and offspring. This correlation exists between relatives as well, for just as parents and offspring are similar to one another, so are siblings and more distant relatives similar to some extent.

Similarity among relatives may be either environmental or genetic, and the two effects may have varying importance in different kinds of species. In humans, in which parents and offspring live together in social groups, much of the similarity results from the family's common environment (i.e., gene—environment correlation). At the other extreme, in a species like a bivalve mollusc in which eggs are released into the sea at an early stage, relatives will not necessarily grow up in similar environments, and shared environments will not exert such a strong influence on similarity among relatives. Darwin's finches probably lie somewhere between these two extremes. It is easy to understand similarity among relatives that is caused by similar environments. In so far as relatives grow up in correlated environments, and environmental variation exists in a character, relatives will be more similar than non-relatives.

The similarity between relatives due to their shared genes is evolutionarily more important. It is possible to deduce this type of correlation among any two classes of relatives from the variance terms we have already defined. We will consider only one case—the correlation between parents and offspring—to see how it occurs. To simplify matters, we will assume that the environments of parent and offspring are not correlated. This assumption allows us to ignore environmental effects (because any environmental effect in the parent will not appear in the offspring—if the parent is larger than average because it benefited from a good food supply, that does not mean its offspring will share those benefits). Any correlation between parent and offspring will then be due to their genetic effects.

The genetic value of the character in the parent, we have seen, consists of several components of which only the additive component is inherited by the offspring. When mating is random, half of that additive component of the individual parent is diluted. At a locus, a parent has an additive deviation from the population mean in both its genes. When an offspring is formed, one of the parental genes goes into the offspring together with another gene drawn at random from the population (we are assuming random mating). The average value of the character in the offspring is, as we saw earlier, half of the additive value of the parent ($^1/_2 A$); the average genetic value in the parent is $A + D$. The correlation between parent and offspring is the covariance between the two:

$$\text{cov}_{OP} = \Sigma \frac{1}{2} A(A + D)$$

where all offspring-parent pairs are summed. The covariance can also be expressed as:

$$\text{cov}_{OP} = \frac{1}{2} \Sigma A^2 + \frac{1}{2} \Sigma AD$$

$^1/_2 \Sigma AD = 0$ because A and D have been defined to be uncorrelated. That is, if an individual has a large value of A, we know nothing about whether its value of D will be large or small. Thus, the equation can be written as

$$\text{cov}_{OP} = \frac{1}{2} \Sigma A^2 + \frac{1}{2} V_A$$

To describe this relation in words, the covariance of an offspring and one of its parents is equal to half the additive genetic variance of the character in the population.

The expression for the covariance between one parent and its offspring is true for each parent. It requires only a small step (though we will not go into it here) to show that the same expression ($^1/_2 V_A$) also gives the covariance between offspring and the mid-parental value. Other expressions can be deduced, by similar arguments) for the covariance between other classes of relatives. The formulae are useful for estimating the additive variances of real characters. However, the estimates become most interesting, for the evolutionary biologist, when expressed in terms of the statistic called *heritability*.

Phenotypic Variance

The similarity between relatives in general, and between parents and offspring in particular, is governed by the additive genetic variance of the character. If a character has no additive genetic variance in a population, it will not be passed on from parent to offspring. For instance, many of the properties of an individual phenotype are accidentally acquired characters, such as cuts, scrapes, and wounds; if we measure these in parent and offspring they will show no correlation. Thus, $V_A = 0$ for such characters. Moreover, some characters, such as the number of legs per individual in a natural population of, for example, zebra, show practically no variation of any sort; for those characters, V_A trivially is zero. Additive variance is, therefore, often discussed as a fraction of total phenotypic variance. This fraction is called the *heritability* (h^2) of a character.

$$h^2 = \frac{V_A}{V_P}$$

Heritability is a number between zero and one. If heritability is one, all variance of the character is genetic and additive. Given that $V_P = V_E + V_A + V_D$, all terms appearing on the right other than *VA* must then be zero. In so far as the factors other than- additive variance account for the variance of a character, heritability is less than one.

Heritability has an intuitive meaning. Consider a parent that differs from the population by a certain amount. If its offspring also deviate by the same amount, heritability is one; if the offspring have the same mean as the population, heritability is zero; if the offspring deviate from the mean in the same direction as their parents but to a lesser extent, heritability is between zero and one. Heritability, therefore, is the quantitative extent to which offspring resemble their parents, relative to the population mean.

How can we estimate the heritability of a real character? One method is to cross two pure lines. This approach is mainly of interest in applied genetics, where the problem might be to breed a new variety of crops; it holds little interest for the evolutionary biologist. The two other main methods involve measuring the correlation between relatives and the response to artificial selection, respectively.

Figure 7.1 is an example that uses the correlation between relatives to determine heritability. The slope of the graph, which shows the beak size in offspring finches in relation to the average beak size of the two parents, is equal to the heritability of beak size in that population. The reason is as follows. The slope of the line is the regression of offspring beak size on mid-parental beak size. The regression of any variable y on another variable x = cov_{xy}/var_x. The covariance of offspring and mid-parental value = $^1/_2 V_A$ and the variance of the mid-parental beak size is equal to $^1/_2 V_p$. (The variance is half of the total population variance because *two* parents have been drawn from the population and their values averaged; if Figure 7.1 had the value for one parent on the x-axis, its variance would be V_p). The regression slope simply equals V_A/V_p, which is the character's heritability. For beak depth in *Geospiza fortis* on Daphne Major, the regression and therefore heritability is 0.79.

Artificial Selection

How can quantitative genetics be applied to understand evolution? We will consider two of the many ways possible here: directional selection and stabilizing selection. A third kind of selection—disruptive selection—will not be discussed here. This section focuses on *directional selection*, which has been studied extensively through artificial selection experiments. Artificial selection is important in applied genetics, as it provides the means of improving agricultural stock and crops.

To increase the value of a character by artificial selection, we can use any of a variety of selection regimes. One simple form is truncation selection, in which the selector picks out all individuals whose value of the character under selection is greater than a threshold value, and uses them to breed the next generation. What will be the value of the character in the offspring generation? Let us define *S* as the mean deviation of the selected parents from the mean for the parental population; *S* is also called the selection differential. The response to selection (*R*) is the difference between the offspring population mean and the parental population mean. In this case, calculating the response to selection is found by regressing the character value in the offspring on that in the parents, where the parents are the individuals that were selected to breed. We plot the offspring's deviation against the parental deviation from the population average to produce a graph. The slope of the graph for parents and offspring is symbolized

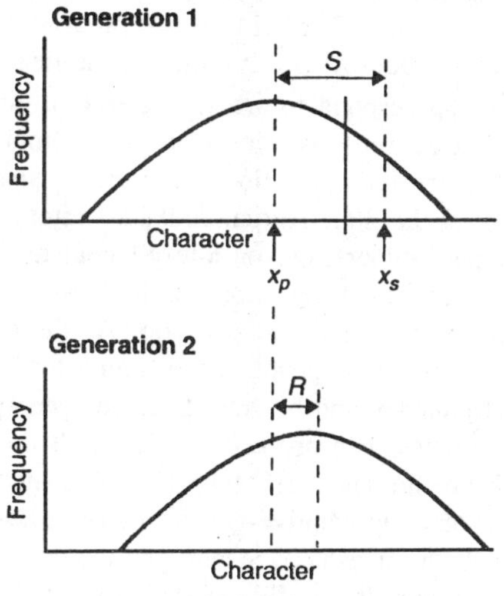

Fig. 7.6. Truncation selection.

by b_{OP} and, as we saw in the previous section, for any character b_{OP} $= h^2$; that is, the parent-offspring regressional slope equals the heritability. Therefore

$$R = b_{OP}S$$

or

$$R = h^2 S$$

This result is important because it shows that the response to selection is equal to the amount by which the parents of the offspring generation deviate from the mean for their population multiplied by the character's heritability. (The response to selection or the parent-offspring regression can be used to estimate the heritability of a character; for a selected population, they are two ways of looking at the same set of measurements.)

A real example of directional selection may not take the form of truncation selection. In truncation selection, all individuals above a certain value for the character breed and all individuals below do not breed. All selected individuals contribute equally to the next generation. It is possible that there could be no sharp cut off, and that individuals with higher values of the character contribute increasing numbers of offspring to the next generation. However, the same formula for evolutionary response works for all forms of directional selection. The difference between the mean character value in the whole population and in those individuals that actually contribute to the next generation (if necessary, weighted by the number of offspring they contribute) is the "selection differential." This value can be plugged into the formula to find the expected value of the character in the next generation.

A population can respond to artificial selection only for as long as the directional selection lasts. Consider, for example, the longest-running controlled artificial selection experiment. Since 1896, corn has been selected, at the State Agricultural Laboratory in Illinois, for (among other things) either high or low oil content. Even after 76 generations the response to selection for high oil content was not exhausted. However, it will eventually stop. As the corn is selected for increased oil content, the genotypes encoding for high oil content will increase in frequency and be substituted for genotypes for lower oil content. This process can proceed only until all the individuals in the population have the same genotype for oil content. At the loci controlling oil content, no additive genetic variance will then be left; heritability will have been reduced to zero and the response to artificial selection will terminate. In the Illinois corn experiment, the process has not yet run its full course. The heritability of oil content in both

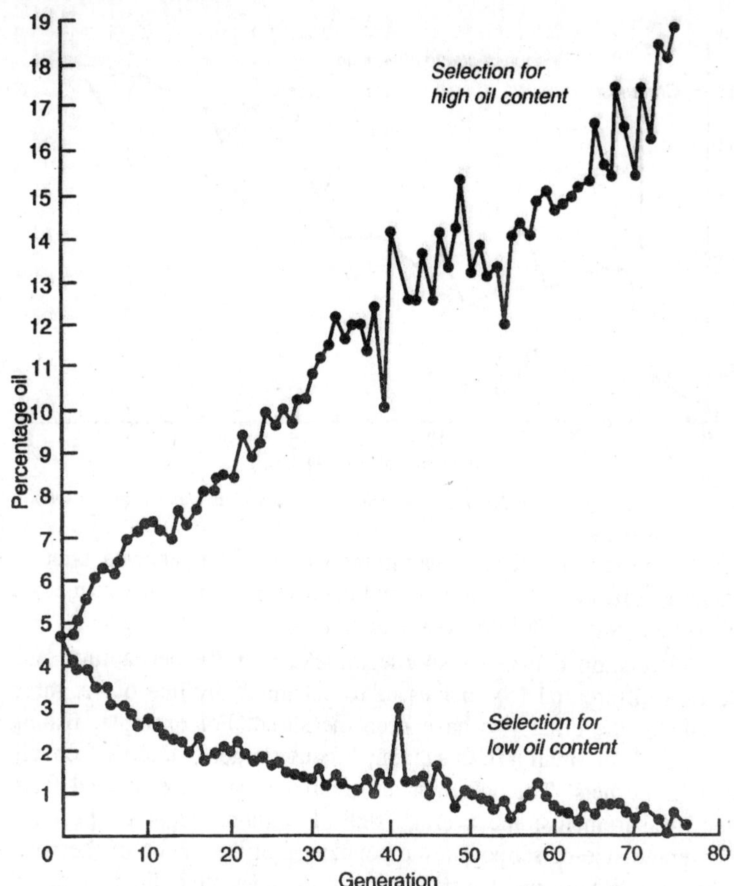

Fig. 7.7. Response of corn (Zea mays) artificially selected for high or low oil content.

the high and low selected lines has decreased through the experiment, but it remains above zero—which is why the population continues to respond to selection.

In other artificial selection experiments, the complete process has been recorded. Figure 7.8 shows the response of a population of fruitflies to consistent directional selection for increased numbers of scutellar chaetae (i.e., bristles on a dorsal region of the thorax). Initially the population responded. Then, as the additive genetic variation was used up (or, as its heritability declined), the rate of change slowed to a stop in generations 4-14. It also appeared that, if selection was continued after the response stopped, the population suddenly began to respond

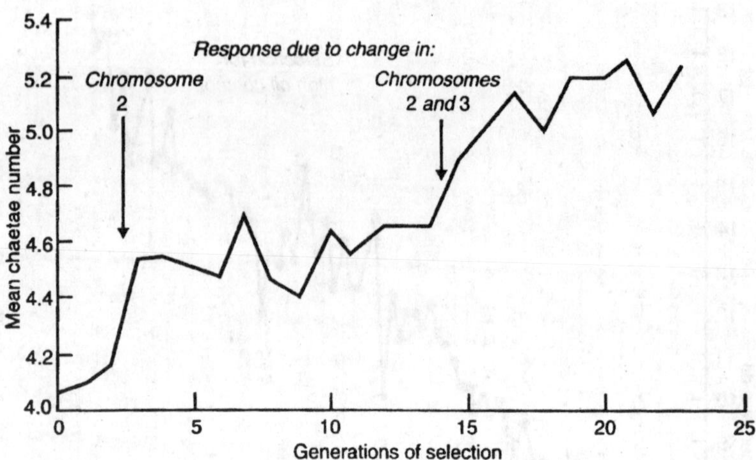

Fig. 7.8. Artificial selection for increased numbers of scutellar chaetae in Drosophila melanogaster.

again after an interval (in generations 14-17). The renewed bout of change is attributed to a rare recombinant or mutation that reinjected new genetic variation into the population.

The relation between response to selection (*R*), heritability, and selection differential (*S*) enables us to calculate any one of the three variables if the other two have been measured. For example, fishing has selected for small size in salmon, because larger fish are selectively taken in the nets. The selection differential S can be estimated from three measurements: the average size of salmon caught in the nets, the average size of salmon in the population at the mouth of the river (before it is fished), and the proportion of the population that is removed by fishing. All three have been measured and lead to the estimates that the salmon who survive to spawn are roughly 0.4 lb smaller than the population average. The response (*R*) for this population—the average size of the salmon decreased by about 0.1 lb between each two-year generation. We can, therefore, estimate the heritability, $h^2 = 0.1/0.4 = 0.25$.

In the finches, Grant measured the response to selection (*R*) and heritability for several characters related to body size, and used these values to estimate selection differentials. We saw that in *Geospiza fortis* that heritability of beak size is about 80%. After the bout of selection for large size in 1976-1977, the finches were approximately 4% larger. We can estimate the selection differential as $S = 0.04/0.8 = +5\%$. As the direction of selection reversed from favouring larger

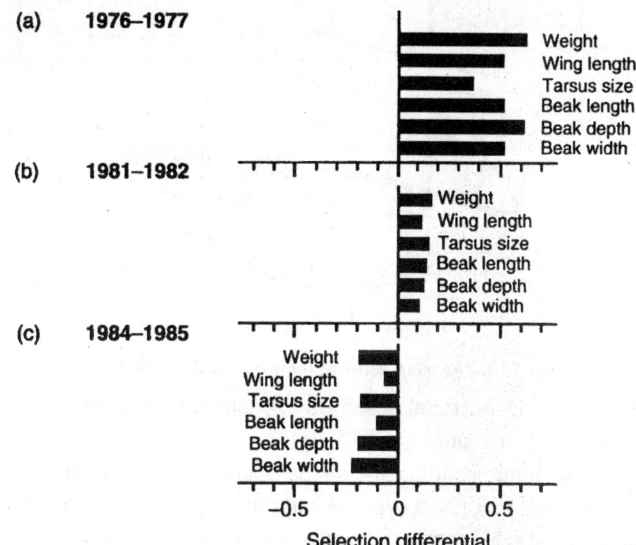

Fig. 7.9. (a) In the drought of 1976-1977, Geospiza fortis individuals with larger beaks survived better and the average size of the finch population increased. (b) In the normal years of 1981-1982, having a larger beak provided a slight advantage, but much smaller than during the drought. (c) After the 1983 El Nino, in 1984-1985, finches with smaller beaks survived better and the average size of the finch population decreased.

beaks in 1976-1978 and smaller beaks in 1983-1985, the selection differentials reversed from positive values in 1976-1977 to negative values in 1984-1985.

In summary, the response to directional selection is controlled by the heritability of a character. The relation between the selection differential, response, and heritability can be used to estimate one of the variables if the other two are known. Sustained selection will exhaust the heritable variation, and the response will then stop.

Relation between Genotype and Phenotype

Scharloo selected a population of fruitflies for increased relative length of the fourth wing vein. The figure shows the frequency distribution of vein lengths in the population for 10 generations. A length of 60-80 has been reached by about generation 5. At this stage the frequency distribution (amid the scatter often encountered in real experiments) starts to show a consistent bimodality. This condition is clearest in generations 5-7, with only the high peak being maintained in generations 8-10. The experiment suggests that more complicated

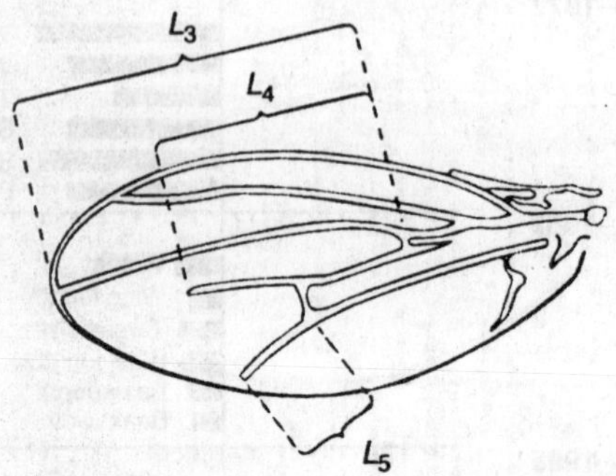

Fig. 7.10. The main veins in the wing of Drosophila.

events can occur in artificial selection experiments than we have seen so far. What is going on?

The key to understanding the shape of the response is the relation between genotype and phenotype. A simple response, such as that for oil content or bristle number, results when an approximately linear relation exists between genotype and phenotype. Genotype here is expressed as a metrical variable. The easiest way to think of this relation is to imagine that the character is controlled by many loci; at each, some alleles (+) cause the phenotypic character to increase, and others (–) cause it to decrease. The more + genes an individual possesses, the higher its genotypic value. When we select for an increase in the character, we pick the individuals with more + genes, enabling the value of the character to increase smoothly between generations in the manner of the Illinois corn oil experiment.

The approximately linear form is not the only possible relation between genotype and phenotype. The bimodal response is thought to result from a threshold relation between genotype and phenotype. The graph has been rotated 180° relative to the form; the *x*-axis (genotype) is drawn down the page on the left. The genotype is believed to control the amount of some vein-inducing substance. Vein length appears across the top of the graph. The relation between substance and vein length is hypothesized to contain a jump at vein length 60-80 (where the artificial selection response becomes bimodal).

Imagine the course of selection for longer wing veins with this threshold relation between genotype and phenotype. The population starts at the top left of the graph with relatively short veins. Initially, some variation appears in the population for genotype, with an associated normal distribution of vein lengths. Artificial selection produces flies with a higher concentration of the vein-inducing substance and with

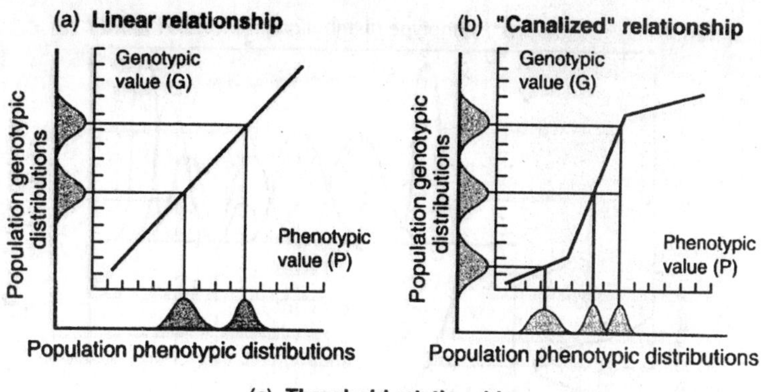

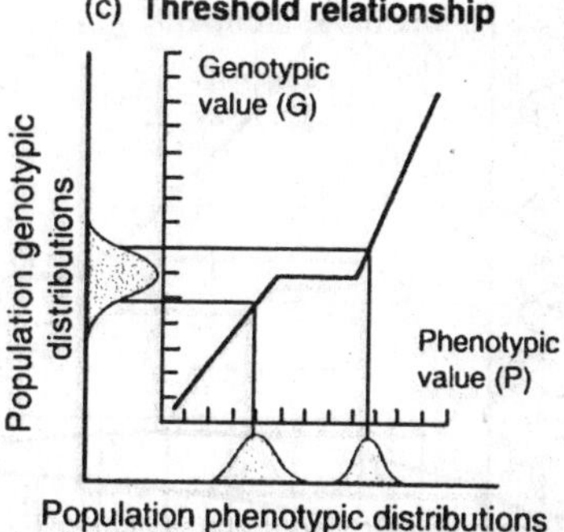

Fig. 7.11. Relation between genotype and phenotype

correspondingly longer wing veins. This process continues until the population reaches a vein length of about 60-80. At that point, the normal (unimodal) variation for the amount of vein-inducing substance generates a bimodal distribution of vein lengths. The bimodal distribution will later disappear, as the population passes beyond the jump in the genotype-phenotype mapping function. Hence, we have the observed response to selection. The relation of genotype and phenotype for vein length is only a hypothesis, but it does show how, in theory, a bimodal response to selection could arise.

The main points to take away from this discussion are that when the genotype-phenotype relation has the linear form, a simple response to artificial selection occurs. The population changes until the genetic variation is exhausted. However, we have no reason to think that this process reflects the typical genotype-phenotype relation. When the relation is more complex, the response to artificial selection can be interestingly different, as the bimodal response to selection on wing vein length in fruitflies illustrates.

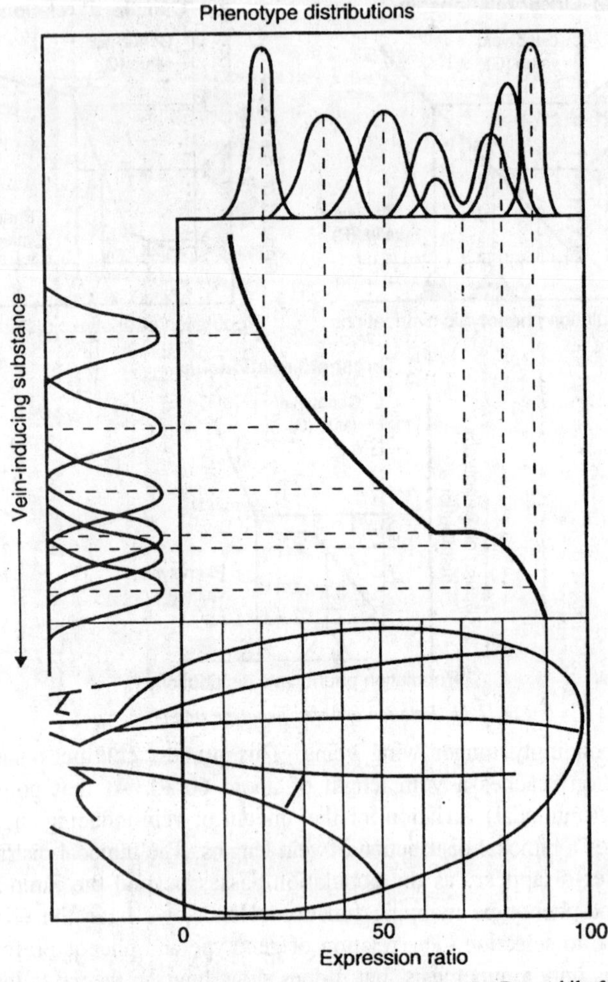

Fig. 7.12. Model of relation between genotype and phenotype for Drosophila fourth wing vein.

Genetic Variability

Natural populations have been undergoing the natural analogy of artificial selection for many generations. They can be thought of as being at the end of, rather than the beginning or anywhere else in, the selection experiment. For characters on which directional selection has been working, the heritability will have been reduced to a low level. The more strongly selection works for a character, the more its heritability should decline, and we might expect that characters closely related to the reproductive fitness of individuals will show low heritability. In any case, directional selection acts to reduce genetic variation for the character.

For many characters, an intermediate rather than an extreme form is probably optimal; the development of this form is called *stabilizing selection*. How much genetic variation should we expect here? Before answering this question, we must distinguish between stabilizing selection and heterozygous advantage. At one locus, heterozygous advantage causes the maintenance of additive genetic variation; if it operated at a number of loci, it would also result in genetic variation in a polygenic system.

However, heterozygous advantage is not synonymous with stabilizing selection. The two concepts differ even in a one-locus system. If, at one locus, *Aa* is intermediate in phenotype between *AA* and *aa*, then any stabilizing selection would coexist with heterozygous advantage. However, if the heterozygote is not intermediate, the identity breaks down. If *AA* is intermediate in phenotype between *Aa* and *aa*, then any stabilizing selection would favour *AA*, not the heterozygote.

The difference between heterozygous advantage and stabilizing selection is more important for polygenic characters. Suppose the value of a character in an organism is influenced by a large number of loci. At each locus two alleles are possible. One increases (+) the value of the character, while the other decreases (-) it. An individual's haplotypes will then each consist of a series of alleles, and might, for example, be symbolized by −++−+−−−++ (for 10 loci). An intermediate phenotype could be brought about by any genotype made up of half + genes and half-genes, such as any of the following:

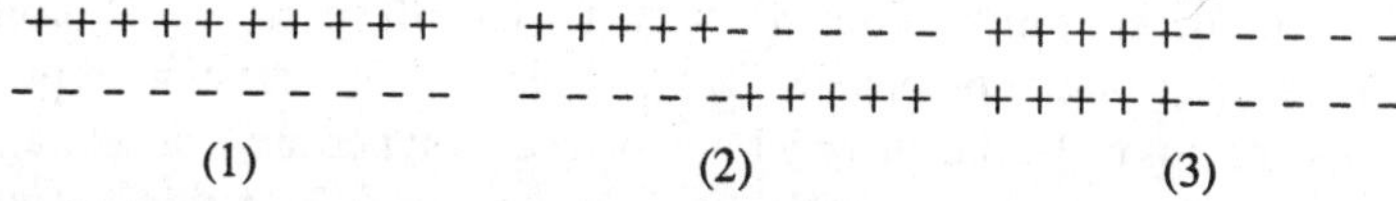

Cases (1) and (2) would produce heterozygous advantage. Selection could also favour the intermediate genotype (3), however, and in a population containing only this haplotype there is no heterozygous advantage.

How much genetic variability do we expect for a character subject to stabilizing selection? In a population containing the four haplotypes, the three genotypes (1), (2), and (3) will all have the same fitness, and we might expect a population to retain considerable genetic variation as these genotypes interbreed and produce a variety of offspring types. However, over a long period of time, genotypes like (3) that breed true will show an advantage, because all of this genotype's offspring have the optimal phenotype, whereas some of the offspring of genotypes (1) and (2) do not. In a population consisting of these three genotypes, selection slightly favours genotype (3). If the environment were constant for a long time, always favouring the same phenotype, selection should eventually produce a uniform population with a genotype like (3).

We can take the argument a stage further. Genotype (3) is not the only true-breeding homozygote that can produce the intermediate optimal form. All of the following also have this characteristic:

+-+-+-+-+-	++--++--+-	+++---++--
+-+-+-+-+-	++--++--+-	+++---++--
(4)	(5)	(6)

Suppose we have a population made up of genotypes (3)-(6), and selection favours the intermediate phenotype. What will happen now? Evolution will again tend toward a population with only one genotype, and that genotype should be a multiple homozygote. The reason behind this result is that any one of the homozygotes that has a slightly higher frequency than the others will have an advantage. Suppose, for example, that genotype (4) had a higher frequency than (3), (5), and (6). All genotypes will now be most likely to mate with a genotype (4). When genotype (4) mates with genotype (4), their offspring will all have the favoured phenotype, identical to their parents. On the other hand, when genotypes (3), (5), and (6) mate with genotype (4), the offspring contain potentially disadvantageous genotypes. The offspring of a mating between a genotype (3) and a genotype (4), for example, will be +-+-+-+-+- / ++--++--+- and has the favoured intermediate phenotype. However, its offspring will contain disadvantageous recombinants. The end result is for selection to produce a uniform population, in which minority genotypes are selected against because they do not fit with the majority form. Selection eventually reduces the genetic variability to zero, even with stabilizing selection.

The argument leads to the following conclusion. If we start with a set of genotypes that influence a character subject to stabilizing selection, and if the environment remains constant for a long enough period, then selection ought eventually to produce a population containing only one, multiply homozygous, genotype.

Stabilizing Selection

The conclusion of the previous section is contradicted by observable facts. Heritabilities can be measured for real characters, and many show significant genetic variation. It suggests that typical values for heritability occur in the range 0.1-0.5. These estimates are for lab populations, and most reliable heritability estimates are unfortunately artificial. Some measurements have been taken from nature as well, such as Grant's measurements, for the Galapagos finch, and these natural observations agree with the conclusion—real characters often have heritabilities higher than zero.

If stabilizing selection tends to eliminate heritable variation, why does it exist? The question is currently a matter of theoretical controversy, and we will look at two possible explanations: mutation-selection balance, and selective processes (like heterozygous advantage) that maintain variation.

Selection-Mutation Balance

Some genetic variation exists in a population as mutation creates new deleterious genes and selection eliminates them. For any one locus, the amount of variation maintained by this selection-mutation balance is low because mutation rates are low. For a polygenic character, however, mutation rates should be approximately multiplied by the number of loci influencing the character. A character controlled by 500 loci will have 500× the mutation rate of a one-locus character. The amount of variability that can be maintained is proportionally increased.

How much genetic variation will exist? This question has been approached from two directions. One approach, which was revived and developed by Lande in the 1970s, considers stabilizing selection on a continuous character (such as body size) controlled by many loci. Mutations at any of the loci can influence the character; because a genotype may be above or below the optimal value for the trait, a small random mutation has a 50% chance of being an improvement. The other approach, which was revived and developed by Kondrashov and Turelli in 1992, does not consider stabilizing selection on a phenotypic character, but supposes mutations occur at many loci and the great majority (many more than 50%) are deleterious. The result is a balance between selection and deleterious mutation at many loci.

We do not have space to go far into either theoretical system, or their relative merits. We will not discuss the other theoretical system at all, but we can look at the experimental evidence measuring mutation rates of deleterious mutation at many loci. The results of this examination have general evolutionary importance.

Deleterious Mutations

The first large-scale experiments to illustrate the destructive power of mutation when selection is minimized were performed by the Japanese geneticist Terumi Mukai. (They were indeed large-scale—Mukai's first experiment in 1964 made use of about 1.7 million flies.) The aim of the experiment was to reduce the effect of selection to a minimum for a number of generations, and measure the decline in fitness. Fitness declines because mutations accumulate—mutations that would normally be eliminated by selection. Mutations can be of varying degrees of destructiveness, from lethal to neutral. The rate of lethal mutations had been measured before Mukai's work, as had the rate of visible mutations (if the effect of a mutation is visible, it is likely to be a large change). These measurements are of only limited interest because lethal and other large mutations probably constitute only a minority of all deleterious mutations. Geneticists suspected the generation of a large class of "slightly deleterious" mutations, which cannot be individually detected at the phenotypic level, and with fitness effects appearing in between the lethal and the neutral mutations. Mukai's experiment revealed the importance of those mutations empirically.

Mukai designed his experiment to transfer a chromosome through a number of generations without recombination or selection, to reveal the way mutations accumulate on it. The experiment used fruitflies (*Drosophila melanogaster*). Fruitfiles have four pairs of chromosomes; they are distinguishable and referred to as chromosomes 1-4. The second chromosome, which was the subject of Mukai's study, makes up about 30% of the total genome. Mukai carried out genetic crosses to create an individual male fly that was homozygous for its entire second chromosome (symbolized +/+): the particular second chromosome came from a fly caught in nature and was not peculiar in any way—the fitness of the homozygotes was in the normal range for homozygotes of natural chromosomes. After creating 104 lines from the original chromosome, Mukai bred the flies through a number of generations, preventing recombination and minimizing selection.

Recombination needed to be eliminated so that the changes in the chromosome would be only due to mutation: if the chromosome could also recombine, some changes would be due to exchange with other versions of the 2nd chromosome. Mukai prevented recombination in the females by means of genetic tricks; no recombination occurs in male fruitflies.

Ideally, Mukai have prevented all selection as well, but that goal is impossible. Instead he minimized it, by two procedures. First, he picked only one male, at random, to father the next generation in each line. Second, the flies were reared in optimal conditions for fly growth, so that as many as possible eggs should have been able to grow up as adults. Selection would not have been reduced to zero, because some of the flies probably had such poor genotypes that they failed to develop; this kind of selection in not excluded in the experiment. Nevertheless, a high proportion of suboptimal flies should have grown up and had a fair chance of being picked to father the next generation. Thus, very little selection operated against slightly deleterious mutations.

As the generations passed, we could have expected mutation to cause the average quality of the flies to decline. Mukai measured the viability (as one component of fitness) of the flies every 10 generations. He sampled individuals from each line and used them to generate flies that were homozygous (+/+) for the chromosome that had accumulated the mutations, or heterozygous for it and a standard lab chromosome (+/standard). The relative survival of the two types of fly were then measured. Any lower survival in the homozygotes relative to the heterozygotes should be due to the accumulated mutations,

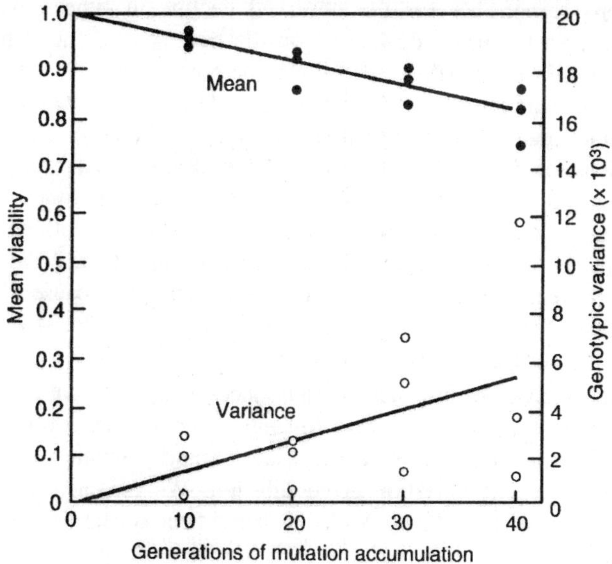

Fig. 7.13. The mutational meltdown in fruitflies protected from selection.

which appear in double dose in the homozygote but only single dose in the heterozygote (allowance has to be made for dominance). In fact, the relative viability of +/+ flies steadily declined.

The result of Mukai's experiment is important for two reasons. First, it reveals the destructive power of mutation. After 40 generations, enough mutations have accumulated in the + chromosome to reduce by 15% the viability of a homozygous +/+ fly relative to a heterozygote. When selection on a population is stopped (or at any rate minimized), the quality of the individuals in it deteriorates over the generations. This fact can be seen even more dramatically, if in less scientific form, by anyone who looks at the flies after a number of generations of the Mukai experience. Mukai himself did not describe his experimental flies, but after 50 generations or so in an similar experiment, the flies are noticeably less vigorous than normal flies.

Second, the quantitative rate of decline in relative fitness can be used to estimate the actual rate of deleterious mutation. We will not work through the full estimation procedure here, except to say that it makes use of the variance in fitness. If the decline in fitness were caused by many mutations of small effect, the variance among lines would be small, whereas the variance would be large if it resulted from a smaller number of mutations of large effect. The actual variance can be used to derive the magnitude, and thus the implied rate, of the mutations responsible for the observed decline in fitness. Several estimates can be made, depending on the treatment of the numbers, but a low estimate is that the second chromosome in Mukai's experiment had a total mutation rate of 0.15 per generation (corresponding to 0.5 per whole genome per generation). That rate applies to all slightly deleterious mutations; thus, each mutation will have a particular effect on fitness, but we can estimate the average effect on fitness of all mutations. The total reduction in fitness, calculated by multiplying the mutation rate and the average fitness effect per mutation, was -0.003 per generation. The number is approximately correct, because after 10 generations the relative fitness of a +/+ fly is about 3-4% lower than at the start of the experiment.

We can conclude that mutation introduces a large amount of genetic variation each generation, and for polygenic characters it is not implausible that selection-mutation balance could maintain quite large amounts of genetic variation at equilibrium. We can also draw one other incidental moral from Mukai's experiment concerning the role of natural selection in the living world. We have met natural selection as a force of change, in peppered moths, pestiferous insects, and

Darwin's finches; we have encountered it as a force for constancy in human birth weight; and we have seen how it leads to the evolution of adaptation. Natural selection is all of these things, but now we have another way of looking at the process. We have just seen what happens to a population in which the effect of selection is minimized. We can imagine by extension how much worse things would be if no selection operated at all. Natural selection in a sense acts to support living things against a ferocious headwind of deleterious mutations that, in the absence of selection, would soon destroy them.

Conclusion

One- and two-locus population genetics is used for characters controlled by one or two loci and whose genetics is known. Quantitative genetics provides the techniques to understand evolution in characters that are influenced by a large number of genes, and for which the exact genotype (or genotypes) producing any given phenotype are unknown. It is possible that the majority of characters have this kind of genetics, in which case quantitative genetics would be appropriate for understanding the majority of evolution. At any rate, it is a highly important set of techniques.

In this chapter, we have seen how quantitative genetics divides up the variation in a character to recognize the component that controls how offspring resemble their parents; the component is called the additive genetic effect. The additive genetic effect plays the same role in quantitative genetics as a knowledge of Mendelian genetics in one- and two-locus population genetics.

The response to selection can be analyzed by means of the heritability of a character, which is the fraction of its variation due to additive genetic effects. However, even with simple directional selection, the exact response depends on the underlying genetic control. For example, the possible threshold relation between the genotype and phenotype for the wing veins of the fruitfly generates an interesting bimodal response to selection. Here, the heritability of the character would show strange changes as the character evolved. Directional selection unambiguously should continue to alter a character until its heritability is reduced to zero. With stabilizing selection, it might be thought that many genotypes could be maintained if they all produce the same intermediate phenotype. However, even here it can be argued that all but one of the genotypes should eventually be eliminated by selection. The argument appears to be contradicted by the facts, and it remains unsettled how mutation and the exact form of stabilizing selection stop populations from becoming genetically uniform.

8

ORIGIN OF GENETIC VARIATION

This chapter is based on the premise that natural selection acted to minimize the deleterious effects of (expressed) genetic errors during the early phases of evolution, when many familiar features of the genetic apparatus were 'fixed'. According to this thesis many key features of the information-transmitting system of cells may be viewed as adaptive responses to the heavy error load placed on early genomes by their inability to correct errors made during copying or introduced from the environment. These features include:

1. diploidy and reciprocal recombination in eukaryotes;
2. genome segmentation among RNA viruses and;
3. the processing system which excises introns from RNA precursors.

The title of this chapter is something of a misnomer. I am less concerned with the origin and nature of genetic variation than with its significance for evolution. Let me start by showing how sloppy use of language unconsciously shapes our ideas, often to the detriment of the facts. Compare the words 'variation' and 'error'. In the context of genetics, both usually refer to mutations in DNA (or RNA), that is, they refer to the same thing. But the mental images they evoke are quite different: variation somehow implies 'useful change'—it links with the notion that mutation is a 'good' thing because it fuels evolution. By contrast error implies 'malfunction'—it links with the idea that mutation is a 'bad' thing because it causes information to decay. This confusion of meanings must be avoided. The hard fact is that although mutations may cause genomes to 'vary' they almost always do so in a *destructive* sense. To use an overworked analogy, when a typist copies a sheet of instructions almost any accidental mistake results in a deterioration of the quality of the information. The chances that a misspelled word or a transposed sentence will result in a more meaningful set of instructions are remote.

Part of the problem is that for two generations most textbooks of biology have enshrined the idea that sexual reproduction and reciprocal recombination have been positively selected because they generate diversity. The implication is that evolution needs more diversity than stochastic processes provide. Although most of us now consciously reject these arguments as 'group selectionist' their influence on our thinking remains deep-rooted and pervasive.

These qualitative statements can be put on a quantitative basis. The error rates in most ancient genes can be estimated by measuring the mutation frequency of polynucleotide copying in the absence of enzymes. The documented values are in the range of 10^{-1} to 10^{-3} substitutions per base per generation. This is to be contrasted with the estimated mutation rate of 10^{-9} to 10^{-11} in modern DNA replication. Simple comparisons like these show that, during the course of evolution, the level of noise in the gene copier mechanism has been slashed by a factor of 100,000,000 or so.

This drop in error levels demonstrates that, at least during much of evolution, natural selection has worked powerfully to decrease the frequency of mutation or to minimize its harmful effects. The best known mechanism of error reduction is correction. That is, secondary channels exist which allow a variety of error-detecting and error-correcting systems to repair premutational lesions in genetic systems. Such corrective processes operate not only during DNA synthesis but afterwards, in the resting state (postreplicative mismatch repair).

Protective Effect of Genetic Redundancy

The present concern is not with error correction but error compensation. The most widespread form of compensation is *redundancy*. The selective advantage, *K*, of redundancy can be modelled mathematically as:

$$K = [1 - (1 - q^L)^n] / q^L$$

where *q* is the base copying fidelity, *n* is the number of copies and *L* is the length of the genetic molecule. The best known example of this protective effect is seen in the structure of the genetic code where the identity of important amino acids is preserved by representing their cognate codons many times over. Fourfold redundancy in the glycine codons, for example, means that 33% of substitution errors in codons will have no effect whatsoever on the phenotype of the encoded polypeptide (assuming that base positions within codons do not affect mutational frequency).

The protective effects of redundancy can be modelled non-mathematically in terms of Figure 8.1. Here we see four DNA molecules encoding two information modules A and B. Provided that there is a critical number of whole genomes, a level of error sufficient to score one 'hit" per molecule per generation should not eliminate intact copies of total information. Thus a reiterated genetic system can survive a degree of genetic scrambling which would consign a single-copy DNA to its doom in short order.

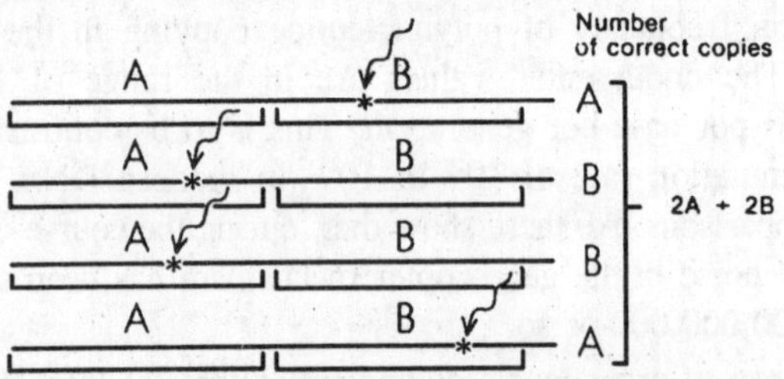

Fig. 8.1. Diagrammatic representation of the protective effect of genetic redundancy.

A striking example of protection due to redundancy may be offered by the unusual prokaryote *Dienococcus radiodurans*. This bacterium is extra- ordinarily resistant to ionizing radiation. This radio-resistance may be linked to the presence of 4—10 'genome equivalents' in each cell of this organism. These DNAs appear to be functionally separate and capable of independent segregation. As we have pointed out elsewhere, in this situation chance works for the preservation of genetic identity following exposure of *D. radiodurans* to ionizing radiation. This is because the *random* nature of mutation ensures that the chances of a given *single* gene being 'hit' in all reiterated copies at *equivalent* positions are remote. The presence of multiple copies of essentially the same information in one 'nucleus' permits recombination between them and this also improves the survival potential of the information since large numbers of multiple crossovers may regenerate some wild-type genomes from damaged ones. I will return to this key point shortly.

Dienococcus radiodurans is adapted to a level of mutation that few other living systems could tolerate. Beyond doubt the most familiar examples of genetic systems protected from error damage by genome redundancy are the *diploid* genomes of eukaryotic cells. It is selectively useful for systems that encode relatively large' amounts of genetic information to present that information to a 'noisy' environment in duplicated form so that mutations in key alleles do not immediately eliminate the organism. The fact that such reiterated systems can store errors as recessive mutations is an 'incidental' consequence of this primary protective strategy. This is not to say that diploidy in modern cells is *maintained* by selection for its protective value. As is often the case in evolution, something that originates in response to one selective pressure may then be subverted to serve quite a separate function. Since diploidy is usually linked to the capacity for sexual reproduction, one can speculate that diploidy arose because it allowed higher cells to alternate between protected (diploid) and unprotected (haploid) phases. Such reversible phases have the capacity to cull out unacceptably damaged genes during cool parts of the seasonal cycle (when the level of beat-induced error is low) and to protect cells during warm intervals (when the level of heat-induced error is high). According to this theory, the familiar alternation of *generations* between haploid and diploid phases had its origins in the equally familiar alternation of *seasons* between cool and hot periods, with the consequent alternation of error rates.

Protective Effect of Genetic Subdivision

Redundancy protects information by *increasing* the mass number of nucleotides per system, i.e. the genetic 'bulk' of the system rises in proportion to the level of protection offered. There is an escalating energy cost to this kind of protection and it is proper to ask if there are alternative mechanisms which allow information to be maintained without incurring such penalties. The answer is 'yes'. It is possible to improve the efficiency of data transfer not by repeating the information but by *subdividing* the information into shorter modules which present a smaller target size to the various error-promoting agents. The principle here is simple: the *shorter* a module of information the *greater* its chances of passing *undamaged* through a noisy channel.

Such divided genomes are today found almost exclusively among the RNA viruses and it can hardly be a coincidence that these tiny genetic objects retain the high error level (3×10^{-4}) characteristic of unrepaired polynucleotide synthesis.

The protective effect offered by genome subdivision can be readily visualized in terms of ultraviolet target theory. Consider an idealized situation in which ultraviolet dosage is calibrated so that it introduces one lethal 'hit' per molecule in a large population of RNAs all of length X. If each RNA is divided into two equal-sized modules, A and

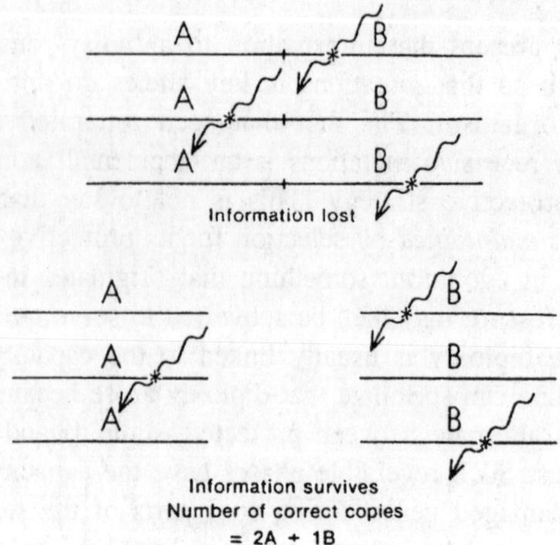

Fig. 8.2. Diagrammatic representation of the protective effect of genetic sub-division.

B, of length X/2 then the same ultraviolet radiation dosage is unlikely to inactivate the whole population. This is because a 'hit' in module A need not affect the integrity of the information in module B, and vice versa. The logic is similar when mutations introduced by an error-prone polymerase are not a result of outside factors.

However, a fundamental distinction must be made between what happens when the modular RNAs are packaged in one capsid (monocompartment viruses) and what happens when each RNA is separately accommodated in a discrete particle (multicompartment viruses). Consider the above example again, and assume that the error rate of the replicase that reproduces the two equal-length modules A

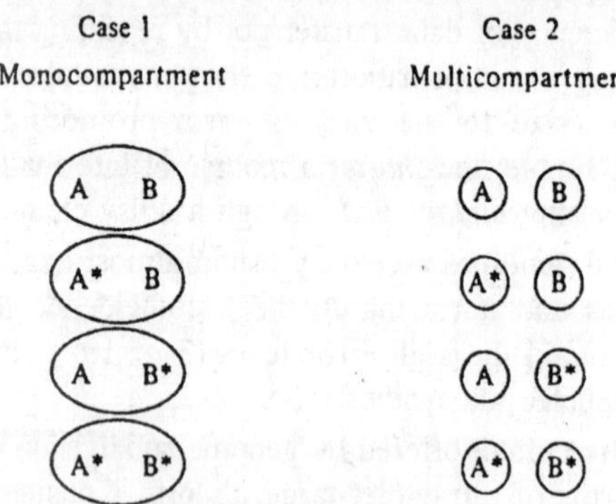

Fig. 8.3. Diagrammatic representation of the protective effect of genetic sub-division in monocompartment and multicompartment viruses.

and B is such that 50% of progeny carry lethal errors. The combinations of modules are then possible, where letters with asterisks indicate lethally damaged modules, and letters without asterisks indicate viable modules. In the first case, each pair of modular RNAs can be accommodated in one capsid so a single particle can initiate infection. However, this strategy limits the number of particles carrying acceptable copies of the genetic information to 25%. In this instance, unless there is some preferential association of non-damaged modules, a segmented genome has no advantage over a non-segmented one. In the second case, 50% of the population of progeny particles carry viable copies of units of the genetic information. This gives the multi-compartment strategy a distinct survival advantage, so long as the transmission of the various encapsidated RNAs remains random and independent. This requirement is met in nature.

The monocompartment viruses appear to have followed an independent evolutionary route. Where multiple RNAs are encapsidated in one coat, sequence-specific interactions must occur if there is to be provision for any kind of discrimination against miscopied information. Does such discrimination exist? Consider the reoviruses. During reovirus maturation *each* particle accumulates the *correct* combination of 10 to 12 *different* modular RNAs. The molecular basis of this specificity is believed to be a set of RNA-RNA or RNA-protein interactions that operate while the RNAs are single stranded.

Lane (1979) has proposed a co-operative process in which the binding for one RNA molecule during assembly alters the nucleation complex to create a binding site for a second RNA and so on. If the binding of a given RNA is faulty, the subsequent RNAs have a lesser chance of entering the nascent particle. Since miscopied RNAs are less likely than faithfully copied RNAs to recognize sequence-specific elements in complementary RNA or RNA binding sites in proteins, their self-amplifying interactions constitute a crude form of molecular proof reading.

Since sequence-specific elements like this are likely to be quite short, one can ask why such interactions—however specific—would select out those molecules whose overall information has been well copied. In answering this question, it is crucial to remember the fundamental difference between a single-stranded and a double-stranded polynucleotide. In double-stranded DNA, the base-pairing capacity of one strand is fully satisfied by its partner, whereas with single-stranded RNA base-pairing capacity is satisfied internally. This means that with single-stranded RNAs *shape* depends on *sequence*. Or, to put it another way, with single-stranded RNA, selection can directly monitor changes in the coding *sequence* through their effects on the three-dimensional *topology* of the molecule. This means that a sequence required to be unpaired (so as to recognize some matching element) must be maintained in its correct configuration by multiple secondary or long-range' tertiary bonds distributed throughout the rest of the molecule. Only if this underlying three-dimensional scaffold is correct can the target sequences adopt the optimal conformation for a faithful interaction either with a complementary sequence on a different RNA or with a stereo-specific RNA binding site in a protein. The target sequence therefore acts as a form of 'quality control' because its availability, i.e. its ability to recognize a matching element, is a sensitive function of the correctness of information elsewhere in the genome.

The reoviruses therefore also conform to my chief thesis in this chapter that genome subdivision is an adaptive response (at least in part) to the high error burden placed on the transmission of information encoded in RNA. Significantly, one can show mathematically that the kind of molecular 'proof reading' envisaged in this model should allow the upper size limit on RNA genomes to be exceeded. It may be no accident that reoviruses, which have the most highly divided genomes of all segmental RNA viruses (10-12 RNA modules), also have the largest genome sizes (12–20 $\times$ 10^6 nucleotides).

'Split' Genes

Sequence-specific interactions among reoviral RNAs provide an intriguing insight into a much wider process. These selective interactions may act as error-screening mechanisms because the reoviral genome is 'split'. But 'split' genes are common in nature. Indeed in the chromosomes of plant and animal cells they are closer to the rule than the exception.

Split genomes and split genes: is there any connexion? I believe there is. Moreover, I believe this connexion may allow us to explain how the exon-intron structure of split genes arose in the first place. The reasoning is as follows.

First, there is now a fairly widespread consensus among biologists that the split gene structure predated the continuous. Perhaps the most compelling evidence for this view is the fact that exons often correspond to modular peptide substructures, which may have discrete functions or roles.

Secondly, a large number of biologists now appear to accept Crick's view that genetic information was originally encoded in single-stranded RNA not double-stranded DNA.

Thirdly, as we have seen, the error rate in the 'first' genes was very high. If we take RNA virus replication as indicative of unrepaired nucleic acid synthesis, we can infer that the error frequency in early genomes was of the order of 10^{-3} to 10^{-4} errors per base per generation. The early copier mechanism was extremely error prone. These postulates together indicate:

(a) early genetic data was encoded in RNA;
(b) such early RNA genomes were 'split';
(c) early replicative processes were very 'noisy'.

This last point is critical. Eigen and Schuster in 1977 worked out a relationship which inversely relates the level of copy error in a self-reproducing system to the maximum size of the information that can be stably transmitted by that system, without experiencing an 'error catastrophe': the *higher* the error rate, the *shorter* the amount of reproducible information. This means that a genetic system without repair capabilities is locked into an 'information freeze'—it cannot evolve towards higher information contents and hence, in an important sense, it cannot evolve at all.

How could genetic systems escape this freeze? Obviously one cannot invoke repair processes such as we see today because these universally use the information specified by the undamaged strand of a duplex molecule to guide restorative processes on its damaged complement. If primitive genetic systems were made of single-stranded RNA it is hard to see how a corrective mechanism that depends on a duplex structure could have evolved or worked.

What options which do not require the *de novo* creation of complex repair mechanisms would be open to such early systems? The most plausible mechanism is genetic redundancy, as this protects the information for the reasons already given, and requires nothing more complicated than the collection of multiple copies of the same information in one 'cell'. Such RNA 'polyploidy' has interesting implications: it immediately allows for recombination between the RNA genomes. Elementary recombination like this need not require protein catalysis in its start-up phases. A model for non-enzymic RNA recombination exists in the self-splicing ribosomal RNA of the ciliate *Tetrahymena*. This transesterification reaction requires strand breakage, strand switching and strand reunion as does DNA recombination, and it occurs in the total absence of proteins. Significantly, the splicing of RNA precursors in the nuclei of higher cells passes through a series of intermediate phases in which branched lariat' structures are produced. There are some tempting similarities between the formation of these branched intermediates and the formation of the postulated intermediates in the auto-excision of RNA from the precursor RNA in *Tetrahymena*.

As Padgett et al. (1984) note of their model for RNA splicing in higher cells: ...this reaction...is consistent with the ribosomal RNA self-catalyzed splicing process seen in *Tetrahymena*. In this latter process, the 3' hydroxyl group from the 5' exon attacks the phosphate group at the 3' splice site in a transesterification reaction to produce the spliced exons and the excised intervening sequence.

I believe then it is entirely reasonable to propose that RNA recombination arose early in evolution for the same reason as DNA recombination did later on. It provided a means of eliminating errors from the pool of mutationally damaged genomes, by regenerating undamaged copies of the genetic information and hence allowing some upward mobility of information content. The selective advantages of recombination between single-stranded RNA molecules are in fact identical to those proposed by Maynard-Smith (1978) for double-stranded DNA recombination. Both function essentially as forms of repair!

If RNA splicing had its origins as a mode of *intermolecular* recombination, what relationship (if any) does it have to the *intramolecular* splicing which occurs today in the nuclei of higher cells? In assessing this question, it is important to bear in mind two points: firstly, the synthesis of RNA from DNA templates retains the high error rate of unrepaired nucleic synthesis; and secondly incorrect splicing has a multiplier or magnifier effect on cell physiology because of its ability to generate out-of-phase messages. In the light of these considerations, it seems likely that the capacity of the modern RNA processing mechanism to discriminate against 'incorrect' sequences is considerable. For example, if errors accumulate in the splice signals or in the sequences that determine topology, then an RNA with multiple exons is unlikely to proceed safely through all the steps needed to remove its introns. This is because RNAs blocked at particular processing steps by mutations in DNA may be degraded in the nucleus. The same is obviously true of errors introduced at equivalent positions during transcriptions. All the processing machinery 'sees' is what is there, whether it came from DNA/DNA or DNA/RNA copying is irrelevant. It follows therefore that the RNA processing machine functions potentially as a kind of molecular 'sieve' or 'filter' ensuring firstly that the average quality of the information reaching the decoding system as message is, in an important sense, *better* than the population of RNA precursors from which it was derived, and secondly that the cell does not waste its energy making nonsense (i.e. out-of-phase) proteins, some of which could damage its internal physiology. The only proviso is that for this system to work, cells must have some mechanism for *separating* primary RNA transcripts and messenger RNAs in a *selective* fashion. It is tempting to ask whether the *nucleus* that characterizes cells with a splicing process was not originally selected for just this reason, as a form of biochemical 'quarantine' to keep in the unproofed primary RNA copies of key genes and let out only those message molecules that had passed the fitness test posed by the multiple-step processing machinery.

According to this thesis, RNA splicing has been retained in the nuclei of cells that encode large' amounts of information because transcriptional copying never evolved efficient repair mechanisms. In DNA, however, the development of a formidable battery of error-detecting and error-correcting processes has seen the error rate fall from about 10^{-3} to 10^{-11} per base per generation. In modern genomes therefore the contribution of genetic noise to the measured rate(s) of mutation is very small. Yet the genomes of higher plants and animals are extraordinarily plastic. Mammalian DNA in particular is something of a genetic museum littered with the 'fossil footprints' of past events due to the insertion and spread of 'selfish' DNA molecules.

Discussion

I believe these two processes—the evolution of DNA editing mechanisms and the development of 'selfish' modular sequences—go hand in hand. Error-correcting mechanisms like the 'cut and patch' process carried out by the *Escherichia coli* DNA polymerase I combine the features of recombination (strand cleavage and ligation) and replication (fill-in synthesis). These kinds of processes could lead rather easily to the kind of 'replicative recombination' that allows semi-autonomous sequences like transposons to multiply in the genome.

This speculation can be put on a more secure footing by going back to Eigen and Schuster's inverse relationship between error rate and genome size. A genetic system operating near the upper limit of its information capacity (as determined by its error frequency) could not tolerate the disruption nor carry the energy burden that the insertion and transposition of large numbers of self-amplifying foreign sequences would bring about. As the error rate falls with improved repair, the maximum information-carrying capacity rises. As genome sizes expand, the cell's energy 'bill' for DNA synthesis also increases. However, beyond a certain point the cell may not, as it were, 'notice' the energy cost of replicating redundant sequences which do not serve any selectively useful function. Such sequences may thus be freed from the selective constraints that dominated previous evolution.

According to this scenario, mutation in modern cells is due less to copy error than to the existence of a multiplicity of mobile genetic elements which have the ability to integrate promiscuously into chromosomal DNA. Evolution in higher plant and animals cells has thus been powered not so much by mutational rewriting of the genetic script as by changes in the 'ecology' of key genes. This idea is consistent with the views that evolution in higher cells is chiefly due to the re-wiring of genetic regulatory circuits and the frequent rearrangements of executive and housekeeping genes.

9

ORIGIN OF GENE

Discoveries of the elaborate structure of genes raise evolutionary questions that lie outside the scope of traditional theory. Darwinism rationalizes how structures evolve, but does not address the more significant issue of the roles that specialized gene structures play in the process of evolution itself. I suggest that molecular genetics can be integrated into evolutionary theory only through the paradigm that underlies all biological explanation; the structure-function principle. Organisms carry out evolutionary processes as biological functions. To do so, species evolve special structures with *evolutionary functions* just as they evolve other phenotypic structures to carry out various adaptive functions. I shall discuss the evolutionary functions of a variety of structures, ways that these functional capacities evolve, and some implications of evolution being a process that organisms become specialized to carry out instead of just change imposed on the species from the external environment.

Heredity lies at the heart of all theories of evolution. Indeed, evolution is the time dimension of genetics. Darwin's greatest handicap in explaining evolution was a complete ignorance of genetics. This forced him to accept two interwoven causal processes in evolutionary change. Natural selection drove evolution by survival of the fittest but an internal hereditary process generated the individual variability on which selection acts. Darwin was unable to evaluate the relative importance of these two agencies. As time went on, he admitted an increasing role for heredity in directing the process of evolution through some sort of Lamarckian mechanism. The discovery of Mendelian genetics rescued natural selection. Recombination of particulate genes, and ultimately blind mutation, produced continuing variation. Since both processes could be completely random and spontaneous, natural selection no longer had to be diluted with organismal hereditary activities. The stark objectivity of the new Darwinism allowed it to overwhelm its vitalistically tainted competitors. Its causal simplicity also bedazzled the 'Neo-Darwinist' into believing that the simplest possible way that evolution could take place must also be the way that evolution actually did proceed.

This misconception might have been exposed by the growing doubts from the fossil record as to whether Neo-Darwinism addresses the key features of the actual history of life on the planet. In fact, it was insight into heredity that again proved pivotal. The revolution of molecular biology has made it possible to examine directly the structure and physiology of genes. Genetic determinants turn out to be very different from the ineffectual abstractions upon which Neo-Darwinism was based. We still cannot foresee their full role in evolution, but it is probably substantial. I suggest that evolution is primarily an internal genetic process. Natural selection, as we know it, occurs and is essential, but it is the junior partner to heredity. Not only the mechanisms but also the goals of evolution are defined by the genetic message. To describe this interpretation I will first sketch the main characteristics of genes as functional structures, then describe the observed evolutionary dynamics of genes, and finally suggest that the phenotype evolves through these same categories of changes.

Nature of Gene

Genes are Units of Molecular Structure

The modern geneticist extracts genes as molecules from cells, looks at them with the electron microscope, measures their physical properties, such as length, and describes their chemical composition. This is entirely different from the approach that prevailed when Neo-Darwinism was being formulated. T. H. Morgan observed in his 1933 Nobel lecture that: 'There is no consensus of opinion amongst geneticists as to what the genes are—whether they are real or purely fictitious—because at the level at which the genetic experiments lie it does not make the slightest difference whether the gene is a hypothetical unit, of whether the gene is a material particle'. Since then, the main concern of genetics has shifted from inheritance of genes to their structure. This approach is fruitful because genes are surprisingly elaborate in structure and this structure is the basis for their function.

Genes are Highly Individual

Until this decade, the quest of genetics was for universal truths about inheritance—Mendel's laws, Fisher's fundamental law of evolution, Crick's central dogma. This made the simplest genes the ones of choice for study. The resulting lowest common denominator orientation reduced genes to no more than runs of codons in DNA molecules, distinctive only in their sequences. Such forms of genes do exist, and probably are abundant, but they seem to be used only for elementary coding functions. Typically, genes with complex information and active evolutionary dynamics are elaborate, and their significant capacities stem directly from their individualistic structure. This realization has a profound evolutionary implication. To understand how important phenotypic traits evolve we must describe the actual dynamics of the complex genes involved, and not merely the stereotyped behaviour of simple genes.

For example, it once seemed all but impossible that the mammalian genome could code for hundreds of millions of different antibody molecules that encompass every possible foreign antigen. Yet vertebrates are able to do so and even make antibodies against artificial molecules to which their species had never been exposed in its entire evolutionary history. Both the capacity to code for the immune response and the ability to evolve this information are made possible only by the very special and unique structure of the immune genes.

Even lowly bacteria rely on highly distinctive gene structures. Although the run of the mill genes in bacteria have a stable configuration on the chromosome, those genes that are actively evolving tend to develop into configurations such as transposons. These structures can move around in the genome and can even hitch-hike from one cell to another by associating with other specialized DNA structures called transmissible plasmids. By exchanging transmissible genes among other species in the ecosystem instead of relying on mutations, bacteria are able to evolve far more rapidly and effectively than would be possible through simple mutation and selection.

It took decades for bacteriologists finally to accept the empirical data that bacteria evolve resistances to antibiotics far too fast for a process of mutation and selection and it has taken evolutionists even longer to appreciate how micro-organisms actually do evolve. As late as 1970, a treatise on evolution still insisted that 'Microorganisms... have the capacity to develop strains resistant to antibiotics and other drugs. This resistance results from the selection of a few resistant mutations of gene combinations exactly as in higher organisms'. With the growth of modern genetics we now know that this is a secondary mechanism for bacterial adaptation. Adaptive evolution in bacteria does not centre on mutation and selection. It is foremost a process of restructuring genes into transposable and transmissible forms and then moving them about the cell and the ecosystem. Thus, for example, the important information for evaluating the potential of a bacterial pathogen to evolve resistance to an antibiotic is not the mutation rate or the frequency of resistant individuals, but the *structure* of the genes that provide that characteristic.

Genes are Working Molecules of the Cell

The third cornerstone of modern genetics is what I call the 'profane' side of the DNA molecule. Classical genetics was predicated on the genotype having a sacred status. Genes were inviolable messages from ancient ancestors passed on faithfully from generation to generation, except for rare mutational corruptions. They were transcendental to the activities and metabolism of the organism carrying them. We now realize that this simply is not true. Instead cells actively manipulate the structure of their DNA molecules for both physiological and evolutionary reasons. They engineer changes in their genes with an impressive array of 'gene-processing enzymes'. Cells seem to have enzymes to catalyze almost every imaginable sort of change in the structure of the DNA molecule. Topoisomerase unties knots in DNA strands, recA protein inserts single strands of DNA into homologous double-stranded helices, acquisitionase grabs bits of DNA and intercalates them into chromosomes and so on. These enzymes for manipulating DNA molecules are not only diverse but also numerous. For example, there are thought to be at least 20 different sorts of DNA glycosylases.

Complex genes, such as multigene families, owe much of their distinctiveness to special enzymic pathways that manipulate their DNA sequences. For example, the antibody and T-cell receptor multigene families have highly specialized enzymic pathways to rearrange their DNA in particular somatic cells and the duplicated genes in the haemoglobin families are enzymically 'corrected' one against another in germ line cells during evolution.

The ultimate authority over the information content of its DNA would be for the cell actually to write gene segments from scratch, and to edit the result in order to achieve a useful product. Remarkably, even this capacity is realized! Mammals are able to polymerize *de novo* sequences of nucleotides at crucial interior sites in the genes for their antibodies. Those physiologically written segments code for particular portions of antibody molecules that recognize their antigens. The creations that are useful are selected from others by an impressively complex multicellular system. A decade ago, geneticists would have considered this capacity unimportant for evolution because it is confined to a small number of highly specialized genes. It is not a general property of genes. However, we now realize that every important gene is probably exceptional in one way or another, and that it is these unique genetic mechanisms that have allowed the important accomplishments of life to evolve.

The full contribution of gene-processing enzymes to evolution remains uncertain. One obvious role is to produce mutants. Probably the majority of mutations important in evolution result from the activities of enzymes. This is significant because enzymes are deliberate and precise, the antithesis of the supposed 'randomness' of mutation. Indeed, enzymically induced changes in the structure of multigene families often are referred to as gene 'switching', 'correction', 'rectification', 'excision', and so on, rather than 'mutation'. I also find it suggestive that geneticists have chosen names for gene-processing enzymes that could be used almost as well for the keys on a word processor.

Genes are Aware of their Surroundings

Genes require information about the organism and its environment in order to express their message appropriately. To meet this demand, complex bacterial genes (operons) have evolved special sensory proteins called 'repressors', which inform them of particular relevant conditions. A typical repressor has two binding sites, one specific for some meaningful metabolite in the cell and a second for a recognition site in the operon. The two binding capacities are mutually exclusive, so that the repressor tells the operon about the concentration of the metabolite by binding to the gene's DNA. Sensor proteins of bacteria are versatile. Some activate rather than repress genes, and some inform genes about hormonal signals generated by the bacterium, or about the status of other genes in the genome. Also, some operons receive information from several sensory systems simultaneously.

Eukaryotic genes are more complex and less well understood than those of bacteria. They are tucked away in a membrane-bound nucleus, and germ line cells are isolated in gonads from the soma. Despite these specializations, receptor proteins still carry hormone molecules directly to sites on the chromosome within the nucleus. Even germinal cells of vertebrates produce receptors on their cell surfaces for a variety of hormones, growth factors, neuromodulators and so forth. Higher organisms presumably have a far greater potential than bacteria for informing their genes but we still know almost nothing of its evolutionary significance.

Types of Evolutionary Change in Genes

In addition to revising our ideas about the fundamental nature of the gene, molecular geneticists also have discovered the types of change that occur in genes during their evolution. Each of these has been chosen as a definition for 'evolution' by one school or another.

Fluctuations in Gene Frequencies

Fifty years ago it was known that genes were located on chromosomes and could mutate to alternative states. Their structure was otherwise mysterious. The only information about genes that could be interpreted (other than map positions) concerned the mathematical *frequencies* of their detectable allelic forms in populations. The obvious basis for a genetic theory of evolution was change in this measurable quantity. Neo-Darwinists therefore defined evolution as simply a change in frequencies of gene alleles in a population from one generation to the next. Such quantitative change is fully reversible for a gene, and the frequencies of individual alleles can shift back and forth to adapt organisms in the population to their changing environment. New alleles occasionally arise by mutation or become 'fixed' by completely replacing their alternative alleles but these processes are very slow and irrelevant to immediate adaptation. In fact, species evolve numerous and diverse mechanisms especially to minimize both the loss of variation by gene fixation and the formation of new mutations, most of which are deleterious. The Neo-Darwinian process concerns fluctuations in the frequencies, but not the forms, of genes.

Change in Gene Structure

When it became technically possible to read the amino acid sequences of primary gene products (and later the nucleotide sequences of the genes themselves), a new paradigm for molecular evolution became the comparison of homologous sequences taken from related species. Species were found to diverge from one another, and from their ancestors, by the accumulation of small substitutions in the molecular structure of their genes. Irreversible permanent change in DNA structure replaced population dynamics as the definition of gene evolution.

'Evolutionary distance' is now measured as the number of substitutions accumulated in the structure of a gene or protein with time. Curiously, this yardstick gives anomalous readings of relatedness among species. It measures the length of time since two species diverged from a common ancestor rather than how much their biology has changed during that time. This is because most base substitutions that accumulate in genes have little effect on phenotype. They are selectively neutral, or nearly so, and driven inexorably into the gene by mutation pressure. Comparative biochemists recognize the distinctiveness of this predominating process of relentless gene change by calling it 'non-Darwinian evolution'. This implies that Darwinian adaptation through natural selection is a separate, and quantitatively lesser, process of change. The main debate in molecular evolution today is about the relative contributions of neutral and adaptive processes to change in gene structure.

Adaptation

There have been two main approaches to the study of adaptive changes in gene structure. One is to expose laboratory populations of micro-organisms to artificial selection pressures and then to analyze the changes by which they adapt. The other is to identify the functionally important parts of genes or proteins and to describe how these vary among related species.

In pioneering the former method, John Langridge spread large numbers of *Escherichia coli* on agar that contained lactobionic acid as the only energy source. This sugar is a derivative of lactose, a disaccharide that *E. coli* metabolizes using the enzyme β-galactosidase. However, the enzyme cannot hydrolyse the analogue sugar effectively. Mutant colonies were readily obtained and genetic analysis showed that many of them had genetic alterations in their β-galactosidase gene.

Langridge's major discovery was the striking variety of *alternative* ways in which the genes were able to acquire this new function. The adapted state is not a unique one, and when two populations or species adapt to a common environment they can be expected to do so in different ways. This raises a key unresolved question. What determines how a gene will evolve when multiple alternative pathways for adaptation are available? The factors that decide the choice between alternatives probably are as important in determining what will evolve as are the environmental stresses to which the species is adapting. So far, few inquiries about gene evolution have reached the degree of sophistication necessary to address this question.

The other approach, of examining the particular parts of macromolecules responsible for functional properties, takes advantage of molecular biology's strong emphasis on structure-function relationships in macromolecules. It is now possible to picture protein molecules in three dimensions and to identify the local regions responsible for substrate binding, catalysis, subunit interaction and so on. Comparing these particular parts of a protein between species shows that adaptation of genes is distinct from just change in structure. Perhaps the most significant revelation from these studies is that the more functionally important a segment of gene is, the slower it changes. In fact, the principal way to locate the functional elements in a gene or protein is to search for regions that have not changed during evolution. This means that selection acts mainly *against change*—a defensive role quite distinct from the Neo-Darwinian view of selection as the driving force *for change* in gene frequencies. Natural selection has been likened to an editor whose main activity is to proofread for errors rather than to promote change.

Advancement

Genes do slowly acquire adaptive changes, but it is clear that not all such changes are of equal importance. Although the majority of selected mutations optimize structural details for the exact functional demands of the environment, a smaller number *advance* the organization of the gene. They create functional capacity instead of adjusting it, and the gene or protein is developed, not just adapted.

Haemoglobin illuminates this process. Most adaptive alterations in this protein probably adjust its oxygen-binding curve to compensate for changes in the organism's size, habitat and so on. In addition, a smaller number of entirely different sorts of changes have *advanced* the haemoglobin molecule from a primitive monomer to its derived tetrameric form. This oxygen-binding protein originally consisted of a single polypeptide chain with a haeme group, a form that still persists in myoglobin. When the chains evolved the capacity to aggregate, they acquired the possibility for allosteric interactions between subunits. This allowed oxygen-loading curves to become sigmoidal with a threshold and a narrowed range between the oxygen tensions for loading and unloading. Allostery also permitted oxygen loading to be regulated by concentrations of other ligands, especially hydrogen ion and diphosphoglycerate. Allosteric regulation is essential to the function of modern haemoglobin, and to enzymes in general. Indeed, nearly all enzymes with sophisticated functions have evolved into oligomers. In the evolutionary history of haemoglobin, the emergence of the capacity for allostery was a landmark advance distinct from its subsequent adaptive fine-tuning.

Several other developments further advanced haemoglobin. One was the evolution of two different types of subunits, alpha and beta, which join together to form the molecule. Heterodimerism allows subunits individually to evolve specialized functions. The importance of this capacity again is evident from the large number of other proteins which also have dissimilar subunits.

Another profound advancement was the organization of the haemoglobin genes into multigene families. Mammals have tandem arrays of genes for the alpha-type and beta-type subunits. The gene copies in these families are differentially expressed during ontogeny to match the haemoglobin molecule to the individual demands of the embryo, foetus and adult. Some species have developed additional haemoglobin gene copies for various other special purposes. Clearly, the vertebrates have actively utilized the opportunities for adaptation opened up by the organization of haemoglobin genes into multigene families. However, the significant evolutionary step was that which created this special adaptability in the first place.

The distinction between advancement and adaptation stands out in many genes, especially in those with demanding biological functions. Significantly, the two processes depend on different kinds of mutations. Genes adapt mainly through base substitutions and occasional small deletions. In contrast, they advance by a wide variety of rearrangements and additions as well as by point mutations. For example, the genes for antibodies acquired their advanced structure by capturing transposons, shuffling exon gene segments, developing special enzymes for rearranging and mutating gene sequences and repeatedly duplicating segments of genes, whole genes and entire multigene families. Molecular geneticists interested in structure-function relationships repeatedly have found it important to distinguish the process that originates and develops structure from the lesser process of its subsequent adaptation. Adaptation obviously plays a role in the origin of novelty, but the two processes are categorically different.

Principles of Evolutionary Advancement

We have a reasonable understanding of how genes fluctuate in allele frequency, change in structure, and adapt. If progression of form is distinct from these processes it is important to determine just how it occurs. Four extra principles seem to be emerging from studies of gene structure. They are discussed here in order of increasing causal complexity.

Tautology of Advancement

Structures that are the most advanced are those that were able to advance the most. This really says no more for evolution than it does for a handful of peas spilled on the ground. Some peas will roll further than others and there must be a reason for the variation. Three can be imagined: (1) some peas might be intrinsically better able to roll, perhaps due to a rounder shape or smoother skin; (2) the surface of the ground may vary from place to place, blocking some peas with bumps and rough spots and facilitating others with depressions or grooves that channel their velocity; (3) chance variation as peas hit the ground and jostle one another will move some peas further than others. The analogous factors in evolution are (1) the genetically determined organization of the species, (2) the selective environment, and (3) whim.

Obviously it is important to evaluate which of these factors have caused some forms of life to advance so much further than others during their 3×10^9 or so years of existence. However, the interactions between phenotype and ecology of higher organisms seem to be vastly too complex for resolving this issue. Fortunately, advancement at the level of gene structure is more tractable. Here, an important observation is that genes that have evolved conspicuously in a particular direction often have identifiable physiological mechanisms that enhance the ability to change in that direction. This association is strong enough for molecular geneticists to rely on it to discover significant physiological mechanisms. For example, a number of multigene families with outstandingly large numbers of gene copies are associated in some way with special mechanisms for reproducing gene copies. These include 5S RNA genes, rRNA genes, pseudogenes, IAP sequences of mice and other retrovirus-like genes, transposons, VAT genes of trypanosomes, possibly satellite DNAs, and unique sequences intercalated into satellite DNA.

As other examples, the extraordinary uniformity of gene copies in some multigene families imply special enzymic mechanisms for removing mutational variation, and one can predict that specialized genes that occur in similar form among unrelated species of bacteria probably are organized as transposons or associated with transmissible plasmids to aid their dissemination. A species also may progress notably in a particular evolutionary direction because it has overcome some factor that constrains such evolution in other species. This represents another way by which a gene or species can specialize its phenotype for a particular kind of evolution. There is a good deal of speculation about structural constraints on the evolution of macromolecules but neither clear examples nor a comprehensive analysis of this issue has yet appeared.

Preadaptation

A particularly important problem for evolutionary advancement concerns the origin of new form. Once a structure advances enough to be functional it can develop further by selection; but how does it initially develop enough to begin to function? This problem worried Darwin and has yet to be resolved. Two principles are probably important. One relates to the well accepted concept of preadaptation. A structure that had evolved originally for one function subsequently acquires a second function. Selection for the original function elaborates the structure until its side effects happened to become useful in other ways. The swim bladder, from which the lung arose, and the reptilian foreleg, which evolved into the avian wing, are familiar preadaptations. The term preadaptation today represents only the opportunism of the evolutionary process without any of the teleological overtones implied by certain earlier evolutionists.

Genes are particularly favourable as preadaptations because they can be duplicated by a single mutation. One copy can preserve the original function, while the other is unhindered to acquire a new function. Moreover, a supernumerary gene can become silent and drift to a new structure unimpeded by selective constraints. Gene duplication and divergence have been dominating processes for evolving new proteins. Most proteins of eukaryotes are members of families that arose from single ancestral genes. These genes of common descent are homologous in the same sense that the wing of the bird is homologous to the leg of the reptile. Of course, not all duplicated genes acquire a new function. The genome of higher organisms is rife with redundant gene copies that have failed to acquire a new function and are degenerating back to random sequence as 'pseudogenes'.

It is significant, but not surprising, that when a duplicated gene acquires a new function that new usage is usually related to the original one. Some groups of homologous enzymes span a wide range of catalytic activities, but most related genes have related functions. This suggests that when genes for a function proliferate, the preadaptive potential for further advances of that general function may increase as well. Some genes have spawned many more new functional derivatives and pseudogenes than others, but it is not clear whether this reflects structural features of the genes or just the functions that they serve.

Autopoiesis

A second mechanism for the origin of novelty involves the physical principle of autopoiesis. This is the spontaneous emergence of organization out of nothing—or, more precisely, out of noise. Closed physical systems and open ones near equilibrium gravitate towards definite simple states of maximal entropy and minimal rate of entropy formation, respectively. Dissipative open systems far from equilibrium have entirely different behaviour. They can spontaneously became more organized with time. A system that is highly dissipative gives fleeting existence to an astronomical number of random or chaotic local microstructures. Any of these twinkling organizational states has the opportunity to stabilize itself from decay and to extend its organization catalytically past its borders if it is able. Just a microscopic seed of such 'autopoietic' organization in an energy dissipating system may grow to macroscopic dimensions, typically as a periodic structure. For example, a thin layer of liquid traversed by a large heat flux characteristically develops a regular periodic pattern of convection that is stable whenever the temperature gradient is applied. Analogous self-forming organization also develops in appropriate dissipative chemical systems and is conjectured to emerge spontaneously even in economic and social systems.

Autopoiesis seems particularly applicable to evolution, which generates new self-propagating organization in a highly dissipative system. A major difference between biological and physical processes is that biological systems are extremely organized to begin with in contrast to the typical system for a physics experiment, such as a container of liquid, on a heat source. The physicist's demonstrations are important to show that dissipating energy does have the capacity to organize matter spontaneously but the appropriate examples to clarify the role of autopoiesis in evolution must come from biology and not thermodynamic paradigms.

Autopoiesis plays an essential generative role in various forms of genetic organization. For example, multigene families easily evolve once they get started because after a gene duplicates it has a propensity to vary further in copy number due to non-reciprocal crossing over. Moreover, the larger a family grows, the more intrinsically variable it becomes. If every additional copy in a multigene family were as difficult to achieve as the first or second, one never would expect to see huge multigene families. However, immense families are common and evolve extraordinarily rapidly. These satellite DNAs give every indication of being products of a runaway autopoietic process.

Autopoiesis has an even more obvious role in the evolution of transposons. A large number of genes scattered along chromosomes and bathed in nucleoplasm rich in enzymes represents the substantial level of organization that characterizes a genome. For example, most cells code for transposases that can synthesize extra copies of a gene and plant them elsewhere in the genome. Brewer's yeast uses a transposase to change sex by copying a silent gene for one or the other sex into a special location where it is expressed. The two sex-determining genes can be copied because they are flanked by special short DNA sequences that the transposase enzyme recognizes as its substrate. Other organisms use transposases for other specialized functions. As long as a transposase, or other gene-processing enzyme, is unrelated to the genes that it acts upon, its activities are causally simple. However, it is possible for the continual shuffling of genes along the DNA molecule to sandwich a transposase gene between recognition sequences for its own enzyme. The enzyme now acts *self-referently* on its own genetic determinant and, in the case of a transposase, a transposon is born. The transposon propagates autonomously by copying itself. Moreover, since its components are genetically specified, the unit can mutate and evolve. Variant copies accumulated in the cell or gene pool can compete with each other for resources, differential fecundity, or *lebensraum* with the 'fitter' form supplanting the others. This evolution is quite distinct from that of other genes because it is *selfish*. Self-reference can allow a gene to evolve for its own sake without dependence on the environment of the organism. Transposons evolve both by modifying the genes they already have and by acquiring new genes.

'Selfish evolution' has been likened to a Darwinian process, with the transposon enjoying the status of an individual 'organism'. Actually the process is considerably more powerful because the 'environment' for the gene is not the inanimate physical world but a milieu determined by a genetic programme. We generally believe that organisms adapt to their environment. However, since transposons can (and actively do) accumulate relevant genes of the host, they are able to adapt their environment to suit themselves. For example, if the amount or specificity of an enzyme in the cell is suboptimal for transposition, a transposon need not adapt itself to that restrictive condition. It can pick up the gene for that characteristic of the 'environment' and selfishly evolve it. Once the gene becomes part of a selfish element it will automatically evolve to optimize its contribution to the fitness of that element.

The transposons that have been characterized (mainly from bacteria) include elaborate forms, attesting to the creative power of their autopoietic evolution. Accumulating such genes does not merely complicate the transposons. It also gives them new emergent properties. Most notably, transposable elements in bacteria have an inherent tendency to become increasingly autonomous of the host. It is believed that genes shuffle actively from one level of autonomy to another as an important process in evolution of bacteria and possibly of eukaryotic genes as well. Thus autopoiesis can create new levels of functional activity and can transfer the determinants of evolution from the external environment to the internal evolving structure.

Evolutionary Function

Some changes in genes are so deliberate that they can only be understood through the biological concept of function. We are well acquainted with adaptive functions, those deliberate biological activities that are carried out to aid survival. In addition, various enzymes that operate on genes have *evolutionary functions*. They do not enhance the fitness or survival of the organism, but instead they facilitate its evolution.

An example of an evolutionary function is the rectification of gene copies in a multigene family. Some families contain dozens or hundreds of gene copies with the same function. It is impossible for the individual copies to evolve directly by natural selection because the redundancy makes the contribution of any one copy totally dispensable. For example, only the most dominantly deleterious mutation in a 5S RNA gene of *Xenopus* could possibly be sensitive to selection with 10,000 sister gene copies in the family. In order to expose sequences in a multigene family to natural selection, organisms must change large groups of the genes to a common structure. If they rectify a block with variant genes to wild type, mutations are purged. If they expand a mutant configuration, natural selection can then favour or eliminate that variant allele of the family within the species.

Some multigene families maintain extraordinary levels of homogeneity. Hundreds or thousands of gene copies are kept almost identical even though they readily evolve in concert. Obviously such uniformity can result only from a very effective rectification mechanism, of which eukaryotes have several. A gene copy can be directly corrected to the sequence of another in some multigene families. Saltatory expansion and contraction, sister chromatid exchange, and extrachromosomal replication and integration probably occur in others. These mechanisms are prerequisites for using multigene families as genetic determinants. Without them not only would a multigene family be unable to evolve its functions, but it would inexorably erode into individual genes. The important point is that rectifying a multigene family promotes its evolution but does not increase fitness. (In fact, the process generates genetic load.) The function of the process is an evolutionary one and not adaptive.

The transposases of bacterial transposons are another set of enzymes with important evolutionary functions. Moving transposons around in a cell does not enhance the cell's fitness at all. In fact, it is probably deleterious on the whole because occasionally a transposon gets inserted into an essential gene. However, transposition is essential to bacterial evolution.

Enzymes are not the only structures with evolutionary functions. Sensory machinery also can regulate cellular processes for evolutionary ends. Repressor systems direct most transposase genes to be active only when appropriate for evolution. Some bacteria and higher organisms have evolved very elaborate sensory systems to induce generalized heritable alterations in their genomes in response to stresses.

Since structures with evolutionary functions have entirely different roles in evolution from those with adaptive functions it is useful to have a specific name for them. Elsewhere I have suggested the term 'evolutionary driver' as the counterpart of the noun 'adaptation'. An evolutionary driver is a genetically determined structure with a function of propelling or steering the evolution of a gene, phenotypic trait or species. Enzyme pathways that rectify multigene families, transposases and their repressor systems, the recA system for inducible mutagenesis in bacteria and so forth, are examples of evolutionary directors which participate in the evolution of genes. If trends continue, we will probably discover a diversity of other evolutionary functions during the next few years. Most of the special organizational features of eukaryotic genes that have surprised geneticists during the past several decades are suspected of having evolutionary functions.

Although the concept of evolutionary function extends some of our ideas about evolution, it retains the conventional notion of 'function'. If the function of the enzyme maltase is to hydrolyse maltose, then surely the function of transposase must be to transpose segments of

DNA, and that of acquisitionase is to intercalate bits of foreign bits of DNA into the chromosome. We would call an effect of a structure its 'function' if it provided part or all of the selective advantage for the development or maintenance of that structure. Organisms can evolve structures with evolutionary functions through the same mechanism of *selection by consequence* that underlies the evolution of adaptive structures. All selection is based on the consequences of a structure rather than the structure itself. Species with greater capacity to adapt to their changing environment will benefit from the consequences of that capacity and can eventually prevail.

As discussed above, preadaptation and autopoiesis greatly increase the ability to evolve new adaptive functional capacities. Most evolutionary drivers probably evolved from adaptive precursors that acted as preadaptations. The repressor on transposon Tn917 is a case in point. Its adaptive function is to repress the erythromycin resistance gene of the transposon when erythromycin is absent from the environment. The repressor also has acquired, probably secondarily, control over the expression of the transposase gene. As a result, the presence of erythromycin induces the erythromycin-resistance transposon to move in the cell in addition to expressing a drug-resistant phenotype. Also, many enzymes used for DNA repair have additional functions in recombination and mutagenesis.

Perhaps the foremost discovery about evolution to come from modern genetics is that organisms carry out evolutionary activities as biological functions. Just how many genes have evolutionary functions is unclear, but the number may be large. Eukaryotes may actually have a greater amount of DNA concerned with evolution than with phenotypic fitness! A wide variety of structures with evolutionary functions should come as no surprise considering the wealth of adaptations that even the simplest species have evolved. Any adapted structure must be considered a possible potential preadaptation for an evolutionary driver. Thus the more a species or gene advances, the more opportunities it has for recruiting adaptive structures into evolutionary roles.

Evolution of Phenotype

As described above, genes change in four distinct ways with time. The frequencies of variant forms fluctuate reversibly in populations, structural alterations become fixed, genes acquire adaptive modifications and they advance in their structure. It is significant that each of these dynamics has its counterpart in the evolution of phenotype.

For example, the distribution of body sizes of European moles fluctuates from one year to another depending on the weather, although there is no evidence for a lasting change. Harsh winters selectively eliminate larger individuals, but milder years allow them to reaccumulate in the population. Presumably body size, like most quantitative traits in animals, is maintained as a long-term balance between intermittent pressures for increase and for decrease. The overwhelming preponderance of phenotypic changes rapid enough to be seen from generation to generation almost certainly are of this nature and leave no mark on evolution over geological time. They are just 'noise' in the long-term process. Most selection on gene frequencies is probably of the stabilizing variety that leads to long-term constancy instead of change in phenotype.

Permanent change in phenotype also does occur, of course. It can be inferred from comparative biology and in some cases can be followed directly in the fossil record. The difficulty is not in detecting change, but in distinguishing changes with no selective importance from adaptations. Obviously changes may be adaptive even though we may not be able to figure out their usefulness. Conversely, and more disconcerting, the selectionist's exercise of devising possible adaptive explanation for a change does not prove that that or any other selective factor caused its evolution. A variety of mechanisms including allometry, mutation pressure, founder effects, fabricational noise, pleiotrophy, drift of various sorts, and gene introgression can cause a phenotypic character to change neutrally or even deleteriously. In recent years, important evolutionists have cautioned against the mistake of assuming that any existing character or phenotypic change automatically must be adaptive. Also, some palaeontologists believe that most morphological change occurs as part of the process of speciation. The role of selection in the formation of new species is poorly understood.

Probably the best guide for picking out the adaptations among evolutionary changes is common sense. The change in the colour of the pepper moth (*Biston betularia*) in England to a darker hue where industrial soot has blackened the tree trunks on which they hide is reasonably inferred to be adaptive. The red colour of blood probably is not. The colour patterns of mollusc shells are more problematical and may include both selective and non-selective factors.

It is easier to document selection for changes over short periods of time than long ones. However, the gulf between the 'ecological time' of the Neo-Darwinist and 'geological time' is so great that there is little reason to believe that the proportion of changes that are adaptive will be the same for the two. It is not even clear which time scale of change is the more constrained by adaptation. One reason for the current interest in the evolution of gene and protein structure is chat it should be so much easier to distinguish adaptive from neutral change where relationships between structure and function are simpler. Yet, even here, selectionists and neutralists bitterly disagree.

Progressive advancement to higher forms is also apparent in phenotypes of organisms. A dominant pattern in the long-term history of life is the appearance of successively higher and higher forms. This does not mean that all lineages inexorably advance. They decidedly do not. The pattern is for most species to get trapped in stasis and for only a minority to continue to advance. Nevertheless, by almost any criterion for 'advancement' that one might choose, life has repeatedly progressed from one record height to another during evolutionary history with the highest forms existing today. Such advancement of phenotype probably involves all four of the principles discussed above for genes.

Tautology of Advancement

The truism that the most advanced forms of life come from the lineages that were able to advance the fastest and farthest suggests that higher organisms evolve by different, and more effective, sorts of changes than do more primitive forms. This is indeed the case. Higher forms evolve in ways that were difficult or impossible for earlier forms, A striking progression has been for life to adapt through modifications first in enzyme catalysis, then in morphology, and ultimately in behaviour. Primitive organisms still evolve largely by changing their biochemical pathways. The main adaptive strategy of bacteria, for example, is to meet changes in their environment with new enzymic activities. To this end they have developed the complex system of transposons and plasmids discussed above. Eukaryotes, in contrast, evolve most notably in their morphology, generating new relationship between cells instead of new types of enzymes. Higher animals continue to evolve the occasional new enzyme pathway, for example, a mammalian lactose synthetase for making milk sugar, but a striking discovery from comparative biochemistry is that *Escherchia coli* has essentially the entire repertoire of basic enzyme activities of human cells. By and large, vertebrates, mammals, man and insects are reflections of special developmental pathways. Morphology offers vastly greater scope for change than do enzymes and, from what we know, is easier and quicker to evolve.

Behaviour allows even more varied and rapid adaptation. The most advanced animals evolve largely through their behaviour. For example, ants are counted among the higher arthropods. Their morphology is not conspicuously advanced but their behaviour makes them outstanding. Ants build large architectural structures, conduct agriculture with both plants and animals, cooperate in societies based on altruism, and take slaves. These functional capabilities emerged from advances in behaviour, although with support from specialized morphology and biochemistry.

Another progression is in the source of variation on which selection can operate. Early forms of life depended on spontaneous mutation. Their advanced descendents developed special enzymic pathways for editing alterations in the DNA. Bacteria evolved the more effective capacity to steal new genes from neighbouring species. Eukaryotes advanced in a different direction, evolving a diploid sexual cycle that generates continual quantitative variation by recombination.

Thus, in various ways, the forms of life that have advanced the farthest in phenotype are commensurately endowed with the greatest capacities for evolving.

Preadaptation

This concept is so familiar in classical evolution that it requires little comment here. Most new functional capacities of higher organisms probably originated opportunistically from pre-existing phenotypic structures.

Autopoiesis

Autopoiesis fits less comfortably with current evolutionary thought because it ascribes the cause of change to endogenous structure instead of interferences from the external world. This is a mode of explanation that Neo-Darwinists have explicitly rejected (although it has gained some recent advocacy from followers of so-called non-equilibrium evolutionary theory). Neo-Darwinism arose in large measure as a repudiation of competing interpretations that evolution was directed from within, such as aristogenesis, orthogenesis, evolution by law, mutationism and so forth. Nevertheless, autopoiesis and self-reference are integral both to the general process of evolution and to specific cases. In fact, the very origin of life was an autopoietic event, with its chicken-and-egg conundrum that faithful replication of the nucleic acid molecule is a prerequisite for the evolution of genetic information, but information for polymerase activity could evolve only after nucleic acids could be faithfully replicated.

Sexual selection is a more specialized type of autopoietic evolution. If female peacocks tend to choose males with large tails as mates, then the genes promoting this attractive masculine character will enjoy a selective advantage. As a consequence, genes that enhance the attraction of females to large-tailed males also will be at an advantage. They will prompt females to mate with the favoured males and hence produce male offspring that are more desirable. Circuits of genes that favour themselves through sexual, social or adaptive processes can erupt into runaway evolution unrelated to economy and ecology. Nevertheless, the directions of such evolution obviously are not random. They must depend in complex ways upon which organizational features of the species are potentially most sensitive to snowballing autopoiesis, and upon the availability of triggers to set the self-reinforcing cycles off. It is suggestive that certain aspects of morphology, such as wing bars, tails and head crests, seem especially prone to exaggeration as sexual adornments of some birds and it is interesting to consider the cause of this pattern.

Far more important self-reference emerges from the action of the organism on its own environment. Neo-Darwinism depends on a Newtonian-like distinction between species on the one hand, and its surroundings as an independent agency on the other. Yet, as Lewontin (1982, 1983) has stressed, this is not realistic. The characteristics of an environment are not independent of the species that inhabit it, but are intimately shaped by the occupants. Probably the most important selective feature of the environment for any higher species is the species itself. Thus, selection pressures are not just dictates of the external world but emerge in large part as the manifestations of the genetic programme of the species itself. This means that the genotype plays two roles in evolution. It responds with adaptive change and it also generates the cause of that change. What a genotype adapts to is largely its own expression. This self-referent cause and effect increasingly dominates evolution as organisms become more competent to manipulate their environments.

Since even the most doctrinaire evolutionist cannot ignore this self-reference completely, Neo-Darwinism has developed special branches of 'frequency-dependent' and 'density-dependent' selection. These have given rise to some really exciting insights into evolution, such as the 'Red Queen paradox' and 'evolutionary stable strategies'. Yet, the most conspicuous feature of deterministic mathematical sojourns into self-reference is their overwhelming complexity. Predictability, optimum and determinism become subordinate to instabilities, alternatives, critical points, feedback and oscillations. If there is one overriding lesson to be learned from the evolution of transposable genes it is that the process of evolution has access to dynamic relationships far beyond the trivia of Newtonian cause and effect. The tautology of advancement implies that significant evolutionary progress is a flag for the more powerful forms of evolutionary cause. I suggest that the important issue is not how the phenotype changes due to the effect of evolution, but how it has become specialized to cause evolution.

Evolutionary Function

An organism's most powerful mode of action is to carry out important activities as biological functions. Recent discoveries about the physiology of genes prove that evolution is able to produce structures to enhance the evolutionary process as well as to aid immediate survival. Much of the phenotype of higher organisms probably functions as evolutionary drivers. For example, the half of eukaryotic organization concerned with sexuality, recombination and mutation did not evolve to enhance the survivability or fecundity of the individual, but for its evolutionary function.

The evolution of sexuality poses a problem for the strict adaptationist because it is a burden on the fitness of the individual. We know that it can be dispensed with because occasional species ranging from plants to insects and vertebrates have become asexual. Scrapping sexuality for asexual reproduction probably aided short-term fitness. Nevertheless, higher organisms have preserved the burden of sexuality except for these sporadic cases because asexuality probably leads to *evolutionary* blind alleys.

Recognizing that phenotypic structures can have evolutionary functions is a separate issue from identifying the ones that actually are evolutionary drivers. I have already discussed the difficulty of sorting out adaptations from selectively neutral changes. Compounding this difficulty is the third possibility of structures evolving to perform evolutionary functions. Even at the level of genotype it is difficult to distinguish among these three possible reasons for a structure's existence. This difficulty is reflected in the fact that geneticists have proposed all three sorts of explanations for each of the organizational features of eukaryotic genes. The problem of assigning functions to structures becomes further complicated when adaptive structures secondarily acquire roles in evolution. They can then serve both sorts of functions simultaneously. An example cited above of a protein with such double function is the repressor of transposon Tn917 which simultaneously regulates both the erythromycin-resistance gene and the transposase gene of the transposon.

A critical task before us now is to assess the extent to which phenotypic traits of higher organisms have acquired evolutionary functions. For example, how widely have canalization systems for reducing noise in development been recruited to guide evolution? A second essential question is how many features of evolution that we take for granted as 'just the way evolution is' actually are results of selection for that mode of evolution? An immediate case is punctuated equilibrium. Is this a specially evolved mode of evolution? Have species developed the capacity or propensity occasionally to reorganize their genomes on a major scale in order to break out of the stasis that otherwise 'straightjackets' established species? Do species evolve special populational, morphological, behavioural or genetic organization to throw off new species? If so, can we explain how and why those features evolved? Also, what about stasis: is the pronounced tendency of many established species to persist with virtually no change an attribute that evolved from selection?

Relationship of Advancement to Adaptation and Change

Both gene structure and phenotype show the four dynamics listed in order of their dependence on each other. Each type depends on change at some underlying level, but may occur with or without producing any higher level dynamics. For example, adaptation presupposes change either in the structure of genes or in the frequencies of gene alleles in the gene pool. However, neither a fluctuation in the composition of the gene pool nor a base substitution in a gene implies that adaptation has taken place.

Advancement is also predicated upon some sort of change occurring but clearly does not inevitably accompany it. Superficially, advancement might seem to require adaptive change but this seems not to be so. For example, coupling a multigene family to an enzyme system that actively corrects one gene copy against another could be a critical advance in the organization of the family even though it had no adaptive importance. Probably complex interactions between the ecology and the structure of organisms determine how fast and in which ways species advance as they change. It is possible that the environment is more important for phenotypic advancement than for advancement of gene structure, but we have yet to understand what situations lead the species to advance instead of simply shifting 'sideways' from one niche to another.

Rates of Evolution

Descriptions of the molecular evolution of genes, and their extrapolation to phenotype, have two implications for rates of evolutionary change. Most notably, *biological structures* play active roles in evolution and must be major determinants of how fast or slow genes and species evolve. For example, their elaborate structures for rearranging and transmitting genes allow bacteria to evolve antibiotic resistance much faster than otherwise would be possible. Genetic and phenotypic organization probably also contribute to slowing down evolution in other cases, although we currently know much less about the mechanisms for stasis. The significance of these genetically determined factors is that they are evolved traits instead of just uncontrolled dictates of the outside environment. To the extent that biological mechanisms participate in evolutionary change they allow its rate to be programmed internally as an attribute of the species.

Secondly, molecular studies show that genes, and presumably phenotype, change in at least four entirely different ways during evolution. These types of change are substantially disconnected from one another and can proceed at independent rates. Analyses of evolutionary rates must begin with a clear description of the sort of change being considered and then proceed to the mechanisms that produce that form of change.

Endogenous versus Exogenous Cause for Evolution

The discovery that genes are inherently dynamic instead of passive structures undermines the basic Darwinian proposition that evolution is change forced on the species by the outside environment. The null hypothesis for Neo-Darwinism is that if a species were boxed into an environment so favourable that it exerted no selection pressure, evolutionary change would cease. This is unrealistic. No matter how much a species was environmentally pampered, its genes would evolve. Species are powerfully influenced by their environment, but environmental forces are superimposed on endogenous engines for change and advancement.

The relative importance of endogenous organization and external environment is perhaps the central question in evolution today. It does not have a simple quantitative answer. One reason is that the species influences its surroundings as discussed above. A more important reason is that the evolutionary role of the genetic message varies from one species to another. In particular, genetic constitution becomes increasingly important as the species advances to a higher form.

Consider the most primitive forms at the dawn of life. Their genes had virtually no capacity to 'do' anything and were maximally exposed to environmental conditions. Before enzyme systems arose to detect and repair lesions and to catalyze change in DNA structure, the mutation spectrum was dictated by physical principles and agents. Evolution probably approached the Darwinian model more closely then than at any time since. Subsequent species evolved increasingly elaborate ways to control mutation, suppressing deleterious types and promoting more productive types. They also developed phenotypic wraps to direct the way that environmental insults forced change on their genes. Higher organisms modified their environment and clustered into aggregates in order to be surrounded by a genetically regulated micro-environment instead of raw nature. Ever more sophisticated adaptations became potential preadaptations for developing ever more sophisticated evolutionary functions. Life asserted greater and greater control over its own evolution as it advanced; the forms progressing the fastest and farthest being those that optimized the factors important to evolutionary advancement. The increasing influence of the species in evolution not only results from advancement but also provides the mechanism for the process of advancement. Evolutionary progress means to gain command over one's evolutionary destiny.

This perspective, that the essence of evolution concerns the capacity to evolve rather than survival of the individual, resolves the enigma of 'higher' forms of life. Theories of evolution restricted to survival of the fittest have to dismiss the whole concept of advancement as illusionary. We might call ourselves and our relatives 'higher' than the protista or prokaryotes but for the survivalist this is just an arbitrary value judgement. By the goal of survival, bacteria are just as fit as we are—perhaps even fitter since they maintain far higher populations and have a phenotype that has survived without significant correction far longer than ours. Certainly we are more *complex*, but is it not just an anthropocentric bias to equate complexity with advancement? Complexity seems to be as irrelevant to fitness as any other general characteristic that varies among extant organisms such as size, life-span, metabolic rate, or degree of specialization.

In contrast, complexity is intimately related to the capacity to evolve. Organization and complexity are measures of both the cause and the effect of evolutionary advancement. An increased capacity to evolve effectively should automatically manifest itself as a more sophisticated phenotype. Reciprocally, phenotypic organization which expands the potential of a species to evolve must advance that process.

This is not just a truism because some phenotypic elaborations can adapt a species exquisitely to a particular environmental factor but impede, instead of facilitate, its subsequent evolution. They decrease evolutionary status even though they make the species more specialized. There is nothing arbitrary or anthropocentric about viewing advancement as a relationship of the organism to its evolution instead of to its external environment. Moreover, it turns the problem of evaluating how advanced a particular species is into an evolutionary question demanding an understanding of evolutionary behaviour, instead of just an ecological issue.

Defining evolution as the process of creating capacities for further evolution has a degree of circularity that is entirely appropriate. The ultimate significance of evolving life is not that it can change and adapt, but that it continually achieves new orders of function. Progressive advancement corresponds more closely to the literal meaning of the word evolution, 'to unfold', than does simple change. What is unfolding, however, is not the organism's fitness for Darwinian survival, but its command over its own evolutionary destiny.

10

GENE MUTATION

Mendel's work in 1900. In 1901 he published his accumulated data in a book entitle. The Mutation Theory. De Varies was careful to distinguish between hereditary and environmental variations, but his mutations are now known to have included several kinds of changes in the hereditary material. It took nearly half a century to clarify the situation in Oenothera and discover the causes of the variations that he observed. Chromosome changes, both structural and numerical, were eventually found to cause the phenotypic variation that de Vries had grouped under the heading of mutations. One complex alteration is described later on in this chapter in connection with the discussion of permanent heterozygotes. It was de Vries' concept of discontinuous variation rather than his precise observations and example that became significant. None of his examples would now be classified as gene mutations.

MUTATIONS AND PHENOTYPIC CHANGE

Because genes are chemical entities that cannot be observed and compared directly, some phenotypic alteration must be associated with the gene change (mutation) or it will go unrecognized. Minor changes presumably occur in genes as a matter of course without producing any phenotypic alterations. Some gene changes are known to be associated with only slight effects on the fertility or viability of the organism. These alterations ordinarily go undetected unless critical comparisons are made. They may, however, influence natural selection and thus represent a factor in evolution.

Mutations also represent in evolution, by which the presence of particular wild-type genes can be postulated. Normal development of any organism is influenced by numerous genes. An observable characteristic of the adult may be altered by an individual gene substitution, but because the entire organism develops as a unit, the action of individual wild-type genes is not always apparent. The existence of mutant and wild-type genes is substantiated when a gene is changed so that its influence on developmental reactions causes a visible difference from that exerted by the unmutated gene. Mutations thus provide the basis for postulating wild-type genes.

Mutated as well unmutated genes tend to give rise to other genes exactly like themselves. Once established, a mutated gene is as stable as the original gene from which the mutation occurred. This process of duplication is associated with mitotic cell division and occurs repeatedly as the number of cells increases. Duplication among genes has a crude parallel in the procedure of printing. After type is set and placed on the printing press; similar copies can be made repeatedly. If the type is changed slightly with an altered template replacing the original, the new pattern reproduces itself faithfully. Instead of a physical change, such as the substitution of one piece of metal for another, the gene in the process of mutation undergoes a chemical change. A part of the gene unit is altered by the chemical modification in such a way that it henceforth behaves differently from its unmutated ancestral gene.

Not all mutations are immediately detectable because many, perhaps the great majority, are recessive and must become homozygous before they can express themselves. Because the phenotypic effect is the only readily observable evidence of mutations, however, a sudden phenotypic change that subsequently proves to be heritable is an accepted indication that mutation has occurred in an organism. In everyday language, the phenotypic change is synomous with the gene change or actual mutation. This is quite natural because the term"mutation" was coined and used before the present gene concept was established. Historically, therefore, the word has been used to describe the perceptible change alteration. It is now known that most mutations that are not detectable by visible phenotypic change influence the viability of the organism in some indirect way.

Classification for Mutations

Dominance and recessiveness provided convenient, although perhaps superficial, criteria for early classifications of mutations. Mutations also may occur in either the autosomes or the X chromosomes, and thus may be classified as autosomal or sex-linked. Viability mutations have also been classified on the basis of their effect on the organism.

Various mutations of this nature in Drosophila have been expressed in the egg, larva, pupa, or adult. When disadvantageous, the mutations are called deleterious; when disastrous to the individual, they are called lethals. Relatively few mutations occur as dominants and even fewer are advantageous in the environments where they occur. The great majority of changes are recessive and detrimental under usual environmental conditions. In a changing environment, a mutation may happen to coincide with new environmental situations and be favourable.

Pleiotropy

A given mutation may alter the organism severely or have such slight effect that it can be detected only when associated with other genes, through cumulative action. The name and gene symbol associated with a mutation are usually taken from the most conspicuous phenotypic alteration that the mutation produces. Thus, white eye (w) is used in contrast to the wild type, red; and ebony (e) body designates the deviation of dark from wild type, gray body colour. This method of designation perpetuates an overly simplified concept of the effects of mutations. Because they exert their influence on basic chemical reactions in the developmental period, single mutant genes affect more than a single trait and may modify in some way every trait of the organism.

The term *pleiotropy* refers to the situation in which a gene is known to influence more than a single trait. For example, in the presence of the white-eye mutation in *Drosphila melanogaster*, the ocelli, malpighian tubules, and testicular envelopes are colourless, whereas these structures in wild flies are coloured. The balancers and wing muscles are altered and the general viability and fertility of the flies are changed. Sheldon Reed has demonstrated a discrimination against white-eyed flies in mating. Such discrimination indicates that the gene causes some effect on the cuticle that influences the mating stimulus. White eye is thus only one several effects of the single mutant gene (w). Many examples of genes with more than single effect have been discovered. All genes (mutants and nonmutants) may be pleiotropic, even though their various effects are not recognized at present. Even though a gene may have many end effects, probably it influences only one primary function in the chemistry of the developing individual.

Congenital hydrocephalus in the mouse, cited by Gruneberg, illustrates the manifold effects of a single gene, ch, on the developing mouse. The gene occurred as a spontaneous mutation and was found by preliminary studies to influence cartilage formation in early development of the mouse. Mice that carried the gene in homozygous condition were born alive but died immediately after birth because the lungs did not inflate properly. Other conspicuous abnormalities also occurred: the skull, forehead, and face were out of proportion; large protuberances filled with fluid extended out from the cerebral hemispheres because the skull bones were abnormal and only skin covered the forehead; eyelids were always open; sensory hairs of the face were abnormal; and the sternum was abnormal with little or no bone formation. Developmental studies showed that abnormalities occurred as early as the thirteenth day in the embryo. The manifold effects were traced back to a single cause, the abnormal cartilage formation. Various skeletal abnormalities were involved directly; physiological disturbances occurred secondarily. A single primary reaction concerned with cartilage formation was controlled by the mutant gene.

Somatic and Germinal Mutation

Mutations may occur in any cell and at any stage in the cell cycle. The immediate effect of the mutation and its ability to produce phenotypic change is determined by its dominance, the type of cell in which it occurs, and the stage in the cycle of the cell. If the mutation occurs in a somatic cell that can reproduce other cells like itself but not the whole organism, the mutant change would be perpetuated only in somatic cells that descended from the original cell in which the mutation occurred.

The "Delicious" apple and the naval orange have resulted from mutations that occurred in somatic tissues. Changes that gave these two fruits their desirable qualities apparently followed spontaneous mutation in single cells, which constituted only a very small part of the body of the individual apple and orange trees that were involved. In each case, the cell carrying the mutant gene gave rise to more of its own kind, eventually producing an entire branch on its respective tree which had the charactristics of the mutant type. Fortunately, it is possible to propagate desirable types of many plants by vegetative methods such as budding and grafting. Vegetative propagation was feasible for both the delicious apple and the navel orange, and today numerous progeny from grafts and buds have perpetuated the original mutation. Descendants of the mutant types are now widespread in apple orchards and orange groves.

Results of somatic mutations have been detected in animals as well as in plants. In Drosophila, white sectors are infrequently observed in an otherwise red eye. Sectors in some male flies have been demonstrated to be composed of descendants from a single cell in which a mutation had occurred (at the w locus in the X chromosome). If the same change had occurred in a germ cell, a white-eyed male might have been produced.

If dominant mutant genes occur in germ cells, their effects may be expressed immediately in the progeny. If they are recessive or hypostatic, their effects may be obscured by other genes. Germinal mutations, like somatic mutations, may occur at any stage in the cycle of the organism but they are more common in some stages, particularly during the gene replication process preceding gametogenesis. If the mutation arises in one of the gametes produced by an individual, a single member of the progeny may receive the mutatant gene. If, on the other hand, a mutation occurs during an earlier stage of gametogenesis, several gametes may receive the mutant gene and therefore several individuals could perpetuate it. In any case, the dominance of the gene and the stage in the cycle when the mutation occurs are major factors in determining the extent of any expression that follows.

Wright noticed a peculiar male lamb with unusually short legs in this flock of sheep. It occurred to him that it would be an advantage to have a whole flock of these short-legged sheep, which could not get over the low stonefences in his New England neighbourhood. Wright used the new short-legged ram for breeding his 15 ewes in the next season. Two of the 15 lambs produced were short-legged. Short-legged sheep were then bred together and a line was developed in which the new trait was expressed in all individuals. The mutation that gave rise to the short-legged sheep was obviously of the germinal type because the cell carrying the mutation had the capacity to reproduce the entire organism. Other examples of germinal mutations have since been described in a wide variety of animals and plants.

Mutation by Deletion or Insertion of Nucleotide Bases

Consider the sequences of bases written in the RNA code shown in Figure 10.1. Now suppose that a base is deleted or an additional base is inserted. The net effect of either a deletion or an insertions that the reading frame is shifted, because different triplets are now read after the mutation site. These are called frameshift mutations. Their net effect is that the amino acids inserted after the deletion or insertion will be different from those in the unmutated cell, and hence the polypeptide will contain a grossly abnormal amino acid sequence

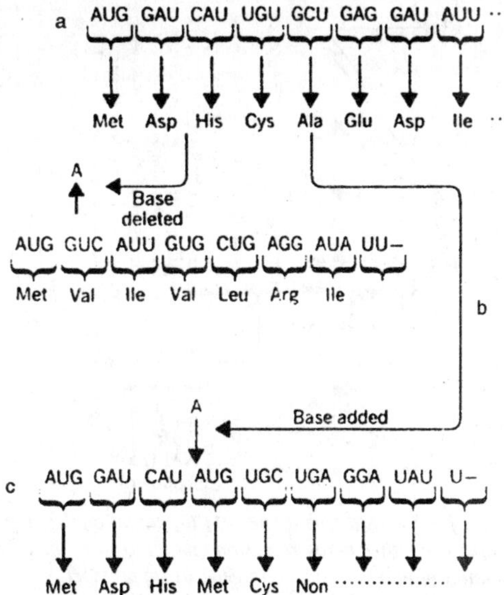

Fig. 10.1. Base deletion and base addition resulting in frameshift mutations. Translation is from left to right.

and be nonfunctional. In addition, somewhere along the line it is possible for a nonsense codon to be produced and the "nonsense" chain will be terminated.

Note that the addition or deletion of two bases will have essentially the same effect as a single base addition or deletion because the reading frame will shift. A three-base change will not shift the reading frame if three immediately adjacent nucleotides are added or deleted, but simply add or delete an amino acid. This is also true for duplication or insertion of adjacent bases in multiples of three. Two or more amino acids will be affected instead of one.

Like nonsense mutations, frameshift mutations can be expected to have drastic effects on the phenotype, unless they occur in multiples of three or cause a change at a nonessential end of a protein. In the latter case, the effect may be great or hardly noticeable, depending again on the position of the deleted or added amino acid(s) in the polypeptide chain. Frameshift mutations can be suppressed by other mutations.

The Reversal of the Mutant Phenotype

So far in this chapter we have been looking at mutation in a somewhat restricted way. It has been tacitly assumed that there is a wild-type allele (or group of wild-type alleles called *isoalleles*, which

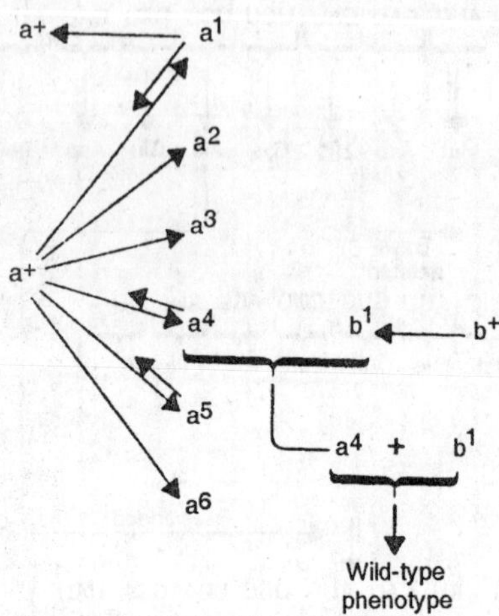

Fig. 10.2. Mutations from a wild-type allele (a^+) to mutant alleles and reversions. The reversions can either be by reversion mutations to a^+ or an isoallele (a^+) or by mutation of another nonallelic gene (a suppressor).

all produce a wild-type phenotype) for each gene that mutates to produce mutant alleles in an unidirectional way. However, mutation can occur either way. The mutant phenotype observed is a result of a forward mutation, and the change back to the original genotypic state is a reverse mutation, by definition. In addition, reversal of phenotype to the original situation, or almost so, may occur without a reversal of genotype but rather by a further mutation in the same gene or in another nonallelic gene. We consider these two situations in this section.

Reverse Mutation

If a missense mutation occurs by substitution of a nucleotide base, it should be expected that it can be reversed by a second substitution in the same codon so as to change it back to the condition in which it will code for the original amino acid, or one which will again give a wild-type phenotype. This has been demonstrated in a most convincing fashion by the use of the chemical mutagens about whose action we have some knowledge. Consider the action of 5-bromouracil, which, pointed out previously, causes transitions of the A:T–G:C type, if 5-bromouracil is able in its enol state to pair with guanine it should also be able to pair with adenine when not in the enol state, as the new chain is forming in the mutant. The net effect is another transition, G:C - A:T, the reverse of the first situation. Therefore, mutations caused by 5-bromouracil should be reversible by 5-bromouracil, and

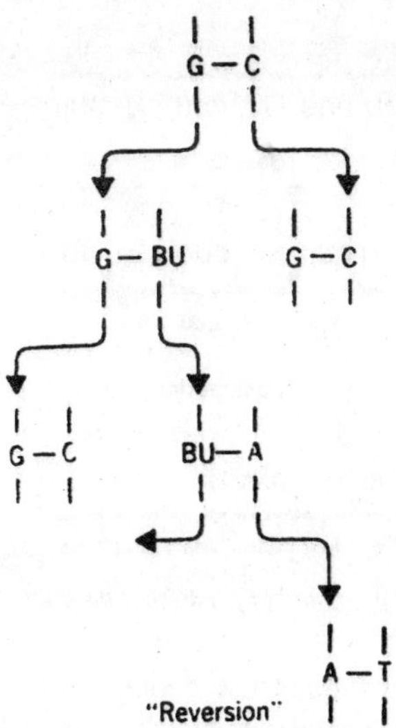

Fig. 10.3. The reversion of a mutation initially caused by 5-bromouracil.

this turns out to be the case. 2-Aminopurine and nitrous acid will also cause reversion of 5-bromouracil-induced mutations.

Hydroxylamine is somewhat different from 5-bromouracil, 2-aminopurine, an nitrous, acid; mutations caused by it are not truly reverted by it. All these results agree with the theory that 5-bromouracil, 2-aminopurine, and nitrous acid should cause transitions in both direction (A:T G:C). Hydroxylamine should cause transitions in only one direction (G:C A:T).

Ethyl ethanesulfonate (EES) apparently causes transitions in the G:C—A:T direction and also transversions such as G:C T:A and G:C C:G. Some, but not all, EES-induced mutants are reverted by base analogues and nitrous acid. The reading frame mutations caused by acridines do not revert in the presence of base analogues or compounds that later bases. And this, of course, is precisely what is to be expected, because acridines cause the deletion and duplication of bases. On the other hand, mutations caused by acridines are for the most part revertible by acridines. In this fact there resides another interesting story; revertants of reading frame mutations may occur in several ways as related in the next several paragraphs.

Intragenic Supressors

An acridine-induced mutant will usually (if not always) result in a gene deficient or duplicated for bases. This shifts the reading frame

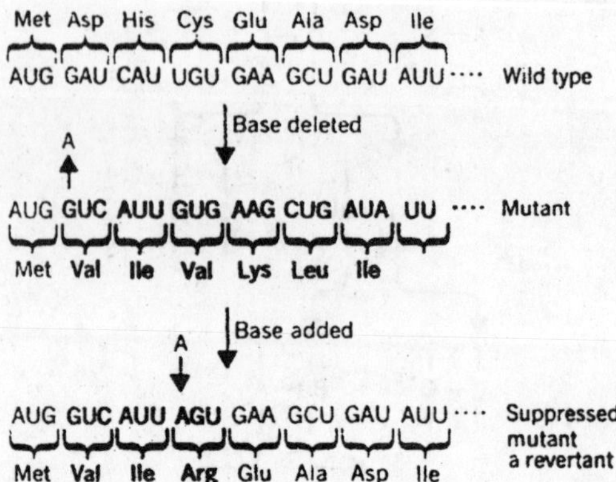

Fig. 10.4. How intragenic suppressors of frame-shift mutations act to produce a functional polypeptide.

and generally results in a nonsense codon, Acridine-induced mutations can be reverted with acridines by the induction of an intragenic suppressor. If the original mutation is a deletion of a base, the suppressor will be an insertion of a base occurring nearby in the DNA. As a result, the reading frame would be brought back to the state in which the proper amino acids are now coded, except for that region between the two mutations. In an analogous way an insertion initially caused by acridine can be suppressed by a deletion. If the amino acids coded for in this region do not form a segment that is important to the protein's ability to function, a true suppression of the mutant phenotype will occur.

A second type of intragenic suppressor involves missense mutations. As stated before, a single amino acid substitution caused by a missense mutation may so disrupt a polypeptide's tertiary conformation that is inactive as a protein. It has been shown that such a mutation may be rectified by a second missense mutation in the same gene. The second amino acid substitution in some way erases the "lethal" change in tertiary conformation caused by the first. This has been demonstrated in the tryptophan synthetase of E. coli and is probably a fairly common even in all organisms.

Extragenic Suppressors

When a nonsense codon is formed either by missense or frameshifts mutation, a drastic phenotypic change is expected because the polypeptide encoded by the affected gene will be incompletely synthesized during translation. Such mutants can be restored to approximately the wild-type phenotype by the occurrence of another mutation in a gene at another locus. Five such suppressors have been described in E. coli. They all suppress one or the other of the two nonsense codons: UAG and UAA or both. It has been shown that in the presence of a suppressor mutation amino acids are actually inserted during translation at points where nonsense codons exist. Hence the polypeptide chains are completed, although the completed chains may not function as well as the polypeptide chains coded in the wild type.

As might be expected, it has been found that the biochemical basis for the action of these suppressor genes is that they code for tRNA's. The sup D gene codes for serine tRNA. In its mutant condition, it produces an altered serine tRNA that introduces serine where a UAG occurs in the messenger. Similar explanations apply for the other suppressor genes, which presumably code for other tRNA's.

These kinds of suppressors have been *supersuppressors*. They are so named because they are capable of suppressing nonsense mutations in a wide range of different genes with quite different and unrelated functions. Theoretically, any gene can mutate by the formation of a nonsense codon. Hence, a single suppressor may effectively suppress specific alleles of a large number of different genes, provided the amino acid it introduces through its altered tRNA results in functional polypeptides. Suppressors have been reported in yeast, *Neurospora*, and perhaps barley. They probably also occur in *Drosophila*. The suppressor of Hairy Wing in *D. melanogaster* suppresses at least 10 other mutant genes, all of which are at different loci and have seemingly unrelated functions.

Suppressors in diploid organisms theoretically should always be dominant, because only one dose of altered tRNA gene should be necessary to produce the tRNA necessary to recognize a nonsense codon. However, suppressors do exist that are expressed only when homozygous. The suppressor of Hairy Wing is one example. Another example in Drosophila is the suppressor of vermilion su(s). It suppresses the mutant condition, sable(s). The action of this type of suppressor, other examples of which are known in *Drosophila*, is not clear.

Missense, Nonsense, and Frameshift Mutations

Various types of mutations can conveniently be discussed in relation to the b-globin gene. It has already been described that the human b-globin gene is split into three coding regions interrupted by two intervening sequences. The nucleotide sequence of the beginning of the first coding region, the sense strand is at the top, and transcription occurs from left to right (from the 3' end of the sense strand to the 5' end). The spaces that separate adjacent triplets of bases do not exist in the actual molecule. It shows the corresponding portion of the b-globin mRNA produced after transcription and RNA processing. Although it is not shown in the figure, the next upstream codon from GUG is the AUG initiation codon, which codes for methionine. Figure shows the first eight amino acids at the amino terminal end of normal b-globin; the initial methionine with which translation begins is not represented in the finished polypeptide because it is enzymatically removed from the chain. The underlined nucleotides represent the restriction site for the restriction enzyme *Mst*II, which cleaves the DNA at this position.

As noted in the previous section, a *base substitution* mutation is one in which a base pair in a DNA duplex is replaced with a different base pair. A base substitution that leads to an amino acid substitution in the corresponding polypeptide is called a missense mutation, and there are two principal types of missense mutations. Part (*a*) again shows a portion of the first coding region of the b-globin gene, and part (B) shows the result of a base substitution in which the normal TA pair indicated is replaced with an AT pair. In this substitution, a pyrimidine base in the sense strant (T) is replaced with a purine base (A), and a purine base in the antisense sense strant (A) is replaced with a pyrimidine base (T). Any base substitution in which a purine (A or G) is replaced with a pyrimidine (T or C), or in which a pyrimidine is replaced with a purine, is called a *transversion*. Transcription and RNA processing of the mutant DNA lead to the mRNA shown in Figure; note that the sixth codon, which normally reads GAG, now reads GUG. Tranlation of this mRNA leads to a substitution of valine for glutamic acid at the sixth position in the polypeptide (black arrow). The transversion mutation is indeed the actual molecular change responsible for b^s hemoglobin. The b^s mutation also obliterates the *MSt* II restriction site, so *MSt* II will not cleave the b^s DNA at this position. This molecular difference provides a rapid and convenient method for *in vitro* diagnosis of sickle cell anemia.

There is another type of missense mutation, which is called a *transition* mutation because it involves the replacement of one purine (in this case G) with another purine (in this case A) or one pyrimidine (in this case C) with another pyrimidine (in this case T). In the transition mutation a CG base pair in the original molecule is replaced with a TA base pair. The corresponding portion of the mRNA is shown in part (C), and in this case the original GAG codon in position 6 is altered to AAG. Translation of the mutant mRNA produces a β-globin polypeptide in which the normal glutamic acid at position 6 is replaced with a lysine. This particular mutation is known as the β^c mutation. Sine the b^c mutation is relatively common in areas of the world where malaria is prevalent, it is thought that heterozygotes for β^c may have some protection against malaria, as is certainly the case with $\beta^{c.}$

Not all base substations are missense mutation. There is also another type of mutation called a nonsense mutation. A nonsense mutation is one that creates a chain-terminating codon [i.e., uAA (orchre), UAG (amber), or UGA (opal)] in a coding region. The name nonsense mutation may seem odd, but it comes from the fact that chain-terminating codons are times referred to as nonsense codons. The transvertion to leads to a UAG terminating codon at the sixth position in the mRNA. Thus, translation of this mRNA leads to a truncated polypeptide because translation terminates at the UAG codon.

In prokaryotes such as *E. coli* and in lower eukaryotes such as yeast, certain mutations have the ability to suppress nonsense mutations. These nonsense-suppressing mutations are called nonsense suppressors, and they involve mutations in genes for transfer RNA (tRNA) that reduce the faithfulness of translation so that chain-terminating codons are sometimes misread.

Such a mutant tRNA will correctly translate serine codons, but, on occasion, will misread a chain-terminating codon (in this case UAG) as a serine codon. Consequently, in some fraction of attempts at translation of the mutant mRNA, the UAG codon will be misread as serine (stipulated arrow), and translation will be able to produce normally beyond this point. If the serine-containing polypeptide is able to function, the phenotypic effects of the original nonsense mutation will have been suppressed by the nonsense suppressor.

Of course, nonsense suppressors can occur in tRNA genes other than serine tRNA. For example, if a UAG suppressor occurs in a gene for glycine tRNA, a UGA codon will sometimes be misread as a glycine codon, and glycine will be inserted into the polypeptide at this position. It should also be emphasized that a particular nonsense suppressor will suppress only one of the chain-terminating codons; that is, an *amber suppressor* will suppress only the UAG (amber) terminator, or *ochre suppressor* will suppress only the UAA (ochre) terminator, and an *opal suppressor* will suppress only the UGA (opal) terminator. Of course, a nonsense suppressor will occasionally misread normal termination codons, too; for example, an amber suppressor will occasionally misread a UAG codon at the end of a normal gene and the result is an abnormally long polypeptide. However, organisms that carry nonsense suppressor are able to survive in spite of the occasional misreading of normal genes.

Not all base-substitution mutations lead to missense or nonsense codons. Because the genetic code contains many synonymous codons, some base substitutions leave the amino acid sequence of the corresponding polypeptide unaltered. Such mutations are said to be silent. In this case a C/G-to-T/A transition at the third position in codon 6 of the β-globin gene leads to an mRNA that has GAA as its sixth codon. However, GAA is a synonymous codon for glutamic acid, so the amino acid sequence of the resulting polypeptide will be completely normal. As a brief inspection of the genetic code will reveal most silent substitutions in coding regions would be expected to involve transitions in the third position of a code. The situation may be very different for intervening sequence, however. Since a particular length and base sequence of an intervening sequence does not seem to be essential for proper gene function, many mutations in intervening sequences—transitions, transversions, even inversions, deletions, or insertions–may be silent.

A final example of a type of mutation that has relatively simple molecular basis is known as a *frameshift* mutation. A frameshift mutation alters the reading frame of an mRNA during translation and thereby greatly alters the amino acid sequence of the corresponding polypeptide. Frameshift mutations are caused by deletions of additions of a small number of nucleotides in a coding region; since the genetic code consists of triplets of nucleotides, any deletion or addition of a number of nucleotides other than an exact multiple of three will cause a shift in reading frame. An example of a frameshift mutation associated with a single nucleotide deletion is Part (a) is again a portion of the first coding region of the human β-globin gene, and the left side of the figure illustrates the normal course of transcription and translation of this sequence. As in previous figure involving this gene, small gaps are introduce in part (c) to show the normal triplet reading frame. The right side of the figure shows the consequence of a deletion of the indicated A/T nucleotide pair. When this deletion-bearing sequence is examined in terms of its triplet reading frame, it can be seen that all triplets beyond the second one are different from those in normal sequence. This lack of correspondence occurs, of course, because the single-nucleotide deletion causes the triplet reading frame beyond the deletion to be shifted one nucleotide to the left. The extensive lack of correspondence due to the frameshift mutation is also evident in the mRNA, and the resulting amino acid sequence in the polypeptide is hardly recognizable as a β-globin sequence. Such out-of-frame translation will continue along the mutant mRNA until a termination codon is encountered.

Mutations Resulting from Unequal Crossing-Over

It has been stated that unequal crossing-over involving duplications can lead to an increase or decrease in the number of copies of the region. When this process occurs at the molecular level, it creates new types of DNA sequences that qualify as mutations. An example of unequal crossing-over creating new mutations is found in the human b-globin gene. Part (a) shows part of the DNA sequence coding for amino acids 83 through 98, which is 18 nucleotides upstream from the second intervening sequence. The boxes indicate two nearby regions that have a perfect eight-nucleotide homology that can act as a duplication. Part (b) shows the coding sequence for amino acids 88 through 103; it has been shifted to the left relative to part (a) to show how the regions of homology can mispair (shaded boxes). As before, the gaps in the DNA sequence indicate the translational reading frame.

Sequence (c) combines the left part of (a) with the right part of (b). This sequence is deleted for the nucleotides that code for amino acids 91 through 95, and it is of some interest that this deletion, called the *Gun Hill deletion*, is found in certain rare individuals. Sequence (d) is the complementary crossover product to sequence (c); this sequence carries a duplication of the nucleotides that code for amino acids 91 through 95.

Mutagenic agents and the Mechanisms of Mutation

Mutations occur in the somatic and germ cell of all organisms. In organisms like Drosophila, mutations are detected for specific loci in one in about 10^5 to 10^6 gametes. Therefore the spontaneous mutation rate for most genes is about 10^{-5} to 10^{-6} mutations per locus per generation. First, agents that increase the mutation rate above the spontaneous level will be discussed because they help explain the spontaneous rate.

The two basic kinds of mutagenic agents are (1) physical ones including radiation (s) and (2) chemical ones, consisting of both "natural" and "unnatural" substances.

The Spontaneous Mutation Rate

An organism carefully protected from known mutagenic chemicals and from known artificially made radioactive sources will still produce gametes or cells with new gene mutations, What causes these spontaneous mutations? A number of factors can be enumerated and may be divided into two categories: environmental and genetic.

Environmental Factors

Background radiation comes to mind as a factor almost immediately, because we have been sensitized since the 1950s to the hazards of ionizing radiation from atomic bombs, industrial and medical applications of nuclear energy, and so forth. However, the average background measured in roentgens per generation is insufficient to account for more than a small fraction of spontaneously occurring sex-linked recessive lethals in Drosophila. But, course, if the background radiation is raised, this could cause additional mutations that might become an important factor in the future of humans and all other organisms.

Ultraviolet light may not be of much significance in the production of mutations since the ultraviolet in sunlight that reaches the surface of the earth has wavelengths only above 300 nm. This range of wavelengths is not effective in inducing mutations to any marked extent except somatic ones in skin. In the higher animals overlying tissue protects the germ cells as already noted.

Chemical in the environment may certainly be a factor, but their effect is difficult to measure for several reasons: (1) a mutagenic chemical may be metabolized to a nonmutagenic state before it reaches the germ cells, or (2) a nonmutagenic chemical may be metabolized to a mutagenic state, and so on. Despite these difficulties, it is now apparent that close attention should be paid to the present and future effects of chemicals on the mutation rate. Our environment is becoming more and more polluted with many different kinds of natural and synthetic chemicals used as insecticides, herbicides, food preservers, medicines, beautifiers, tranquilizers, and hallucinogens. What effect do these how have on the human mutation rate? We do not know, but it is important that we find out, because an increase in the rate could have diastrous results.

For the present, we can assume that the presence of chemicals in the environment may have some effect on the mutation rate of all organisms, but it certainly cannot account for all the spontaneous mutation rate. Metabolic products, such as formaldehyde and hydrogen peroxide produced within cells, may also be involved, but here again it is doubtful that their effects are highly significant.

Genetic Factors

Organisms evolved in a changing environment to which they had to continually adapt as they evolved. Mutations were necessary for this evolution to occur, just as they are necessary for future evolution. But the mutation rate must not be so high that it results in the production of many deleterious mutation that would act as "genetic load" on the population.

It is highly probable that the main component of the spontaneous mutation rate is genetically controlled. As we have described in previous pages, enzymes are involved in the replication of DNA. Mistakes of different types may occur during replication that may lead to mutation, especially if the replicating enzymes themselves are deficient. These mistakes may be replicated or they may be repaired by the repair systems that we known to exist. Thus, two opposing systems, genetically controlled, may be postulated to exist: the replicative synthetic one and the repair system. The spontaneous rate of mutation may be determined by an equilibrium point between them.

It has long been known that different populations of *Drosophila melanogaster* collection from widely scattered areas in different part of the world show significantly different mutation rates (Table 10.1). Furthermore, genes have been identified that have an influence on the mutation rate A second chromosome gene, *hi*, and a third chromosome gene, *mu*, have been located in *melangoaster*. Both of these increase the "spontaneous" mutation rate when homozygous, although *mu* appears to act in this way only in females. Extensive work has been done in E. coli with mutant strains that show a high rate of A:T $\rightarrow$ C:G transversions. The result is that the mutation rates as specific loci are raised several thousand fold!

Genes that raise the mutation rate are called *mutator* genes. A number of these have been identified in yeast, *Drsosophia*, and E. coli. They have yet to be directly implicated in the DNA replicative and repair processes, but it is difficult to believe that they are not.

In addition to mutator genes, there are also numerous examples of mutable loci in a variety of organisms. Certain alleles of some genes exhibit an unstable condition, so that they mutate from the wild type to mutant condition at a high rate in either the germ cells, or somatic cells, or both.

Table 10.1. Spontaneous mutation rates found in different populations of *D. melanogaster*, sex-linked and second chromosome recessive lethals

	X chromosome			*Second chromosome*		
Stock	*Number tested*	*Number lethals*	*Percent lethals*	*Number tested*	*Number lethals*	*Percent lethals*
Florida inbred	2108	23	1.09	—	—	—
Wooster	1266	8	0.63	—	—	—
Oregon R	3049	2	0.07	—	—	—
Florida No. 10	916	10	1.09	516	9	1.74
Lausanne	955	2	0.21	436	3	0.69
Leningrad	8614	14	0.16	—	—	—
Sukhami	2309	24	1.04	—	—	—

Phenotypic Effects of Mutations

Mutations must normally cause some detectable *phenotypic change* for their presence to be recognized. The effects of mutations on phenotype range from alterations so minor that they can be detected only by special genetic or biochemical techniques to gross modifications of morphology to lethals. A gene is a specific sequence of nucleotide pairs coding for a particular polypeptide. Any mutation occurring within a given gene will thus produce a new form or *new allele* of that gene. Because of the degeneracy of the genetic code, some base pair changes do not change the protein products coded for by the genes in any way. Genes containing mutations with small effects that can be recognized only by special techniques are called "isoalleles." Other mutations result in total loss of gene-product activity. If mutations of the latter type occur in essential genes (genes required for viability), they will, of course, be lethal.

Mutations may be either recessive or dominant. In haploid (or, more accurately, monoploid) organisms like viruses and bacteria, both recessive and dominant mutations can be recognized by their effects on the phenotype of the organism in which they originated. The dominance or recessiveness of mutations in bacteria can be determined only by studying partial diploids. In diploid (or polyploid) organisms, recessive mutations will be recognized only when present in the homozygous condition. Most recessive mutations in diploids will not be recognized at the time of their occurrence, since they will be present in the heterozygous state. Sex-linked recessive mutations are an exception, since they will be expressed in the hemizygous state in the heterogametic sex (males in humans and fruit flies; females in birds). Sex-linked recessive lethal mutations will alter the sex ratio, since hemizygous individuals carrying the lethal will not survive.

The most useful mutations for the genetic analyses of many biological processes are conditional lethal mutations. These are mutations that are (1) lethal in one environment, the so-called restrictive conditions, but are (2) viable in a second environment, the permissive conditions. Such mutations allow geneticists to identify and study mutations in essential genes that result in complete loss of gene-product activity even in haploid organisms. Mutants carrying conditional lethal can be propagated under permissive conditions, and information about the functions of the gene products can be deduced by studying the consequences of their absence under the restrictive conditions. Conditional lethal mutations also provide valuable selective mutations also provide valuable selective sieve for genetic fine structure analysis.

The three major classes of mutants with conditional lethal mutations are (1) *auxotrophic mutants*, (2) *temperature-sensitive mutants*, and (3) *suppressor-sensitive mutants*. Auxotrophic mutants (as opposed to prototrophic "wild-types") are mutants that are unable to synthesize an essential metabolite (amino acid, purine, pyrimidine, vitamin, etc.) That is synthesized de novo by wild-type individuals of the species. Such auxotrophic mutants will grow and reproduce when the metabolite is supplied in the medium (the permissive condition); they will not grow when the essential metabolite is absent (the restrictive condition). Temperature-sensitive mutants will grow at one temperature but not at another temperature. Most temperature-sensitive mutants are heat-sensitive some, however, are cold-sensitive. The temperature sensitivity usually results from increases heat or cold lability of the mutant gene product, for example, an enzyme which is active at low temperature but partially or totally inactive at higher temperatures. Occasionally, only the synthesis of the gene product is sensitive to temperature, and once synthesized, the mutant gene product may be as stable as the wild-type gene product, Suppressor-sensitive mutants are viable when a second genetic factor, a suppressor, is present, but are nonviable in the absence of the suppressor. The suppressor gene may correct or compensate for the defect in phenotype that is caused by the suppressor-sensitive mutation, or it may render the gene product, altered by the mutation, nonessential.

Most of the thousands of mutations that have been identified and studied by geneticists have been found to be deleterious and recessive.

This is to be expected, considering what we know about the genetic control of metabolism and the techniques available for identifying mutations. Metabolism occurs by sequences of chemical reactions, each step of which is catalyzed by a specific enzyme coded for by one or more genes. Mutations in these genes frequently produce blocks in metabolic pathway. These blocks occur because changes in the base-pair sequences of genes often (but not always) cause changes in the amino acid sequences of polypeptides, which may, in turn, result in loss of function. This, in fact has been the most commonly observed effect of easily detected mutations. Given a wild-type allele coding for an active enzyme and mutant alleles coding for less active or totally inactive enzymes, it is apparent why most of the observed mutations might be recessive, as is observed. If the enzyme is metabolically important, such mutations will also be deleterious. If the enzyme catalyzes an essential reaction, the mutations causing total loss of activity will be recessive lethals.

But why should most mutations with phenotypically recognizable effects result in decreased gene produce activity or no gene-produce activity? This result can be predicted if one accepts the effectiveness of natural selection and if one accepts the effectiveness of natural selection and if one assumes the existence of a semiconstant environment during the recent (on the evolutionary scale) evolution of life forms on earth. A "wild-type" allele of a gene coding for a "wild-type" enzyme or structural protein will have been selected for optimal activity for many generations. Mutations resulting in amino acid changes that increased the efficiencies of enzymes in carrying out particular functions will have been preserved by natural selection, and, as the most "fit," they will have become the new "wild-types". Given a sufficient period of time in semiconstant environment, most, if not all, sequences of amino acids (at least all sequences of amino acids that differ form the "wild-type" by a single mutation) will have been tried, and natural selection will have preserved the most efficient one. It will now be the wild-type. Mutations, which produce changes in these very specific sequences of amino acids, will usually result in less activity or no activity at all. As such, they will most frequently be recessive and deleterious.

An analogy can be made with any complex, carefully engineered machine. If you randomly modify any one essential component (e.g., of a watch or an automobile), it seldom performs as well as it did prior to the random change.

Radiation-Induced Mutation

That portion of the electromagnetic spectrum containing wavelengths that are shorter and of higher energy than visible light (wavelengths below about 0.1 μm) can be subdivided into ionizing radiation (X rays, gamma rays, and cosmic rays) and nonionizing radiation (ultraviolet light). Ionizing radiations such as X rays (about 0.1 to 1 nm) are of high energy and thus are useful for medical diagnosis because they can penetrate living tissues. In the process of penetrating matter, these high-energy rays collide with atoms and cause there lease of electrons, leaving positively charged free radicals or ions. These ions, in turn, collide with other molecules, causing the release of further electrons. The net result is that a "core" of ions is formed along the track of each high-energy ray as it passes through matter of living tissues. This process of ionization (thus the name ionizing radiation) is induced by machine-produced X rays, protons, and neutrons, as well as by the alpha, beta, and gamma rays released by radioactive isotopes of the elements (e.g., ^{32}P, ^{35}S, radium, cobalt-90 etc.). Ultraviolet rays having lower energy, penetrate only the surface layer of cells in higher plants and animals and do not induce ionization. Ultraviolet rays dissipate their energy to atoms that they encounter, raising the electrons in the outer orbitals to higher energy levels, a state referred to as excitation. Molecules containing atoms in either ionic forms or excited states are chemically more reactive than those containing atoms in their normal stable states. The increased reactivity of atoms present in DNA molecules is the basis of the mutagenic effects of ultraviolet light and ionizing radiation.

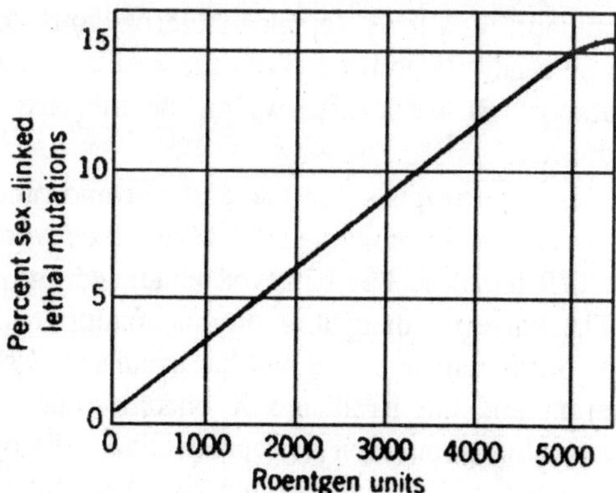

Fig. 10.5. Relationship between the frequency of sex-linked lethal mutations induced in Drosophila sperm and ionizing irradiation dosage.

Ionizing Radiation

In 1927, H. J. Muller first demonstrated that mutation could be induced by an external factor. Muller demonstrated that X-ray treatment markedly increased the frequency of sex-linked recessive lethal mutations in D. melanogaster. Muler's unambiguous demonstration of the mutagenicity of X rays became possible by his development of a technique facilitating the simple and accurate identification of lethal mutations in the X chromosome of Drosophila. This technique, called the ClB method, involves the use of females heterozygous for a normal X chromosome and X chromosome (The ClB chromosome) specifically constructed for Muller's experiment.

The ClB chromosome has three essential components. (1) The C (for crossover "suppressor") refers to the presence of a long inversion, that prevents recombination of genetic markers on the *ClB* chromosome and alleles on the normal X chromosome. The inversion does not actually prevent crossing over, but causes gametes containing X chromosomes produced by crossing over between the *ClB* chromosome and the normal X chromosome to be inviable. Chromosomes resulting from crossing over between a chromosome containing an inversion (an inverted segment of the chromosome) and a normal chromosome will contain duplication and deficiencies (repeated and missing sets of genes). The inversion is required in Muller's experiment to assure that the markers on the *ClB* chromosome stay together through meioses. (2) The *l* refers to a recessive lethal in the *ClB* chromosome. (3) The *B* refers to the presence of the partially dominant mutation that causes the bar eye phenotype, which is a narrow, slit-shaped eye. Because it is partially dominant, it allows females heterozygous for the *ClB* chromosome to be readily identified. Both the recessive lethal (*l*) and the bar eye mutation (*B*) are located within the inverted segment of the *ClB* chromosome.

Given females heterozygous for the *ClB* chromosome. Muller's experiment was operationally quite simple. Males flies were irradiated and mated with *ClB* females. The bar-eyed irradiated and mated with ClB females. The bar-eyed daughters of this mating will carry the *ClB* chromosome of the female parent and the irradiated X chromosome of the male parent and the irradiated X chromosome of the male parent. Since the entire population of reproductive cells of the males was irradiated, each bar-eyed daughter carries a potentially mutated X chromosome. That is, each male is likely to produce some sperm that contain X chromosomes carrying new lethal mutations and some sperm without X-linked lethal mutations. The bar-eyed daughters were then mated individually (in separate bottles) with wild-type males. If the irradiated X chromosomes carried by a bar-eyed daughter contains a sex-linked lethal, all of the progeny of the mating will be female. Since males are hemizygous for the X chromosome, those receiving the *ClB* chromosome will die due to the recessive lethal (*l*) that it carries. Those receiving the irradiated X chromosome will also die if a recessive lethal has been induced in it. Matings of bar-eyed daughters carrying an irradiated X chromosome, in which no lethal mutation has bee induced, with wild-type males will produce female and male progeny in a ratio of 2:1 (only the males with the *ClB* chromosome will die). Scoring for the presence of recessive sex-linked lethals is thus unambiguous and error free using the *ClB* technique—simply scoring for the presence or absence of male progeny. By using this technique. Muller was able to demonstrate an increase in mutation rate of up to 150-fold after X-ray treatment.

Another technique that facilitates the detection of mutations in Drosophila, in this case sex-linked mutations with visible effects on phenotype (often called "visible mutations"), involves the use of *attached-X-chromosomes*. Attached-X-chromosomes undergo compulsory nondisjunction (failure of homologous chromosomes or chromatids to disjoin or separate during anaphase) since the two X chromosomes are joined to single centromere. If females with two attached-X chromosomes plus a Y chromosome (XXY) are mated to normal males, any mutation that occurs on the X chromosome of the male will be expressed in the surviving male progeny. In such attached-X matings, the male progeny receive their X chromosome from their male parent, rather that from their female parent as in a normal mating. If the male parent is treated with a mutagenic agent such as X rays, the increased frequency of recessive visible mutations can be easily assessed by screening the male progeny of the attached-X mating.

X rays and most other forms of ionizing radiation are quantitated in roentgen units (r units, pronounced "runtgen"), which are measured in terms of the number of ionizations per unit volume under a standard set of conditions. More specifically, one roentgen unit is the quantity of ionizing radiation that produces one electrostatic unit of charge in a one cm^3 volume. Note that the dosage of irradiation in roentgen units does not involve a time scale. The same dosage may be obtained by a low intensity of irradiation over a long period of time or a high intensity of irradiation for a short period of time. This is very important because in most studies the frequency of induced point mutations is directly proportional to the dosage of irradiation. In Drosophila sperm, for example, there is an increase of approximately 3 percent in the mutation rate for each 1000 r in increase in irradiation dosage. This linear relationship between mutation frequency and dosage in indicative of so called "single-hit-kinetics." That is, only one event (ionization?) or one "hit" is required to cause a mutation. Or, stated differently, every ionization has some fixed (under a specific set of conditions) probability of inducing a mutation. If the cumulative effects of many ionizations were required to induce a mutation, would plot as a curve that was concave upward.

The linear relationship between mutation rate and radiation dosage is important because it speaks directly to the frequency asked question of What is a safe level of irradiation? Even very low levels of irradiation have certain low, but very real, probabilities of inducing mutations. The question is thus meaningless. There is, no such thing as a safe level. In *Drosophila* sperm, for example, very low levels of irradiation over long periods of time (chronic irradiation) are as effective in inducing mutations as the same total dosage of irradiation administered at high intensity for short periods of time (acute irradiation). This clearly has major practical significance in evaluating the effects of the increased exposure of living organisms to radiation that results from the testing and use of nuclear weapons and nuclear reactors in generators, spaceships, and so on.

In mice, chronic irradiation has been found to induce somewhat fewer mutations than the same dosage of acute irradiation. Moreover, when mice were treated with intermittene doses of irradiation, the mutation frequency was slightly lower than when they were treated with the same total amount of irradiation in a continuous dose. It should be emphasized that all of these irrdiation treatments were mutagenic, albeit, to different degrees, to both Drosophila and mice. The different responses of fruit flies and mice to chronic irradiation may result from differences in their ability to repair damaged DNA. Repair mechanisms may exist in the spermatogonia and oocytes of mice that do not exist in Drosophila sperm.

The single-hit theory implies that one ionization can produce one mutation, but it does not imply anything about the efficiency with which this happens. Several factors have been shown to affect the efficiency with which irradiation induces mutations. The receptivity of a cell at different stages of its metabolic cycle is an important factor in determining rates for induced mutations. A. H. Sparrow has shown marked variations in numbers of chromosome fragments, assumed to be directly related to mutational changes, at different stages in meiosis and cleavage in the plant genus Trillium. Chromosome aberrations were induced about 60 times more frequently at metaphase than it interphase. Nondividing cells in Trillium showed little radiation damage, whereas rapidly dividing cells were very sensitive.

Oxygen tension and temperature change, when associated with irradiation, also may significantly alter the frequency of mutations. Low-oxygen tension decreases mutations. Oxygen can magnify the effect of radiation, but only if it is present during the irradiation. Oxygen has less effect with intense conditions than with moderate conditions of ionization. Environmental agents that protect germ cells from radiation damage often do so by lowering the oxygen concentration of tissue, and those that enhance the effectiveness of radiation and oxygen.

Ionizing radiation also induces various kinds of gross changes in chromosome structure (chromosome aberrations) such as deletions, duplications, inversions, and translocations. These changes in chromosomes structure result from breaks in chromosomes caused by ionizing radiation. Since they require two breaks, the kinetics of induction are two "hit" as expected, rather than single "hit."

Ultraviolet Radiation

Ultraviolet rays do not possess sufficient energy to induce ionizations. They are, however, readily absorbed by certain substances such as purines and pyrimidines, which then enter a more reactive or excited state. Because of their lower energy, they penetrate tissues only slightly, usually only the surface layer of cells in multicullular organisms. Nevertheless, ultraviolet light (UV) is at a wavelength of 254 nm. Maximum mutagenicity also occurs at 254 nm, suggesting that the UV-induced mutation process is mediated directly by the absorption of UV by purines and pyrimidines. In vitro studies show that the pyrimidines (especially thymine) absorb strongly at 254 nm and, as a result, become very reactive.

The two major products of UV absorption by pyrimidines appear to be pyrimidine hydrate and pyrimidine dimers. Several lines of evidence indicate that thymine dimerization is probably the major mutagenic effect of UV. Thymine dimers appear to cause mutations indirectly in two ways. (1) Dimers apparently perturb the DNA double helix and interfere with accurate DNA replication. (2) Occasional errors are made during the processes that cells possess for the repair of "damaged" DNA, such as DNA containing thymine dimers. The relationship between mutation rate and UV dosage is highly variable, depending on the type of mutation, the organism, and the conditions employed. "single-hit kinetics" are only occasionally observed, in contrast to ionizing radiation.

Mutation Frequency

Mutation is necessary to provide the genetic variability required for the evolutionary adaptation of species to environmental changes. On the other hand, most are deleterious. Thus, if mutation were to become too frequent in a species, it would create a sizable "genetic load" of deleterious effects. Clearly, if this "genetic load" became too large, the species would face extinction. Human technology has already contaminated the earth with increased levels of radiation and chemicals that are known to be mutagenic. While the consequences of the present increased level of mutagenes in the environment cannot yet be accurately assessed, most scientists agree that further increases in the levels of mutagens in the environment should be avoided. Yet significant quantities of hundreds of new chemicals are introduced into the environment each year, most of them with insufficient, if any, mutagenicity tests. Whether they will cause harmful increases in mutation rate (and/or cancer incidence) is a question of utmost concern to everyone.

Each gene probably has its own characteristic mutational behaviour. Some genes mutational behaviour. Some genes undergo mutations more frequently than others in the some organism. Those with unusually high mutation rates are called unstable or mutable, but a wide range of mutation rates exists among genes that are considered stable. The mutation rate per gene in bacteria is of the order of 1 in 100,000 to 1 in 10 million (10^{-5} to 10^{-7}) per cell generation. For fruit flies, the average for mutation in a particular gene is in the order of 1 in 100,000, with a range from 1 in 200,000 to 1 in 100,000, with a range from 1 in 20,000 to 1 in 200,000 gametes. Since most of the data on mutation rates in fruit flies have been obtained from experiments with males, questions have arisen concerning a possible sex difference in overall mutation rates. B. Wallace has shown through extensive experiments that mutation rates are not significantly different in the two sexes of D. Melanogaster. However, mutation rates do differ in different strains.

Estimates of mutation rates humans indicate a somewhat greater frequency than those cited for stable gens in most other organisms.

Samples collected thus far have been small, and the methods used were indirect and subject to large errors. Genes associated with such human traits as intestinal polyposis and muscular dystrophy have been estimated to mutate once in 104 to 105 people. A human generation is equal to about 50 cell generations. By expressing any mutation rate as probability of mutation per cell per generation, a mutation rate is defined independently of exact physiological conditions and stage in life cycle. This definition is based on a time unit proportional to a cell's division time. When expressed in terms of cell generations, rates for fruit and humans are generally comparable with those for bacteria.

Estimating Mutation Rates

Mutation rates are usually defined in terms of mutation events per generations or, less frequently mutations per unit time. The direct experimental determination of mutation rates is complex. Because mutation is must be sampled to get accurate estimates of mutation rates. In addition, recurrent sampling is required unless the initial population is sufficiently small that its probability of containing a mutant organism is negligible. Simply enumerating the number of mutant individuals in a population does not indicate how many mutational events have occurred since the mutants will usually undergo exponential growth like the nonmutants, although often at a slower rate than the latter.

Consider a "wild-type" allele a^+ mutating to a mutant allele a with a mutation rate μ (defined as a constant a probability of a mutational event per cell duplication or per organism duplication), that is,

$$a^+ \rightarrow a$$

clearly, the number of mutation events will depend on the number of a+ alleles in the population. For simplicity, consider a haploid organism. If one starts with population in which the number of mutant organisms equals *M* and the total population size equals *N,* then

$$dM = \left(\mu + \frac{M}{M}\right)dN$$

This *assumes* that (1) M is very small relative to N, so that the difference between N and N – M is negligible; (2) the mutant and nonmutant organisms duplicate at the same rate; (3) reversion of a → a+ does not occur or is negligible. If one starts with a population in which M is zero, and makes all the same assumptions, then the relationship simplifies to

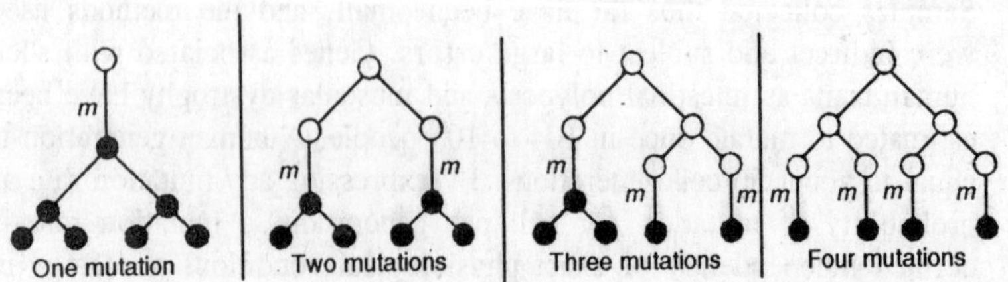

Fig. 10.6. A given number of mutant organisms in a population can result from various numbers of mutational events, depending on when the mutation occurs during the exponential growth of the population.

$$dM = \mu dN$$

These equations can be integerated and used to obtain estimates of mutation rates in experimental populations. More frequently, however, populations are analyzed in terms of allele frequencies, or changes in allele frequencies due to mutation, selection, or migration. Moreover, mutation is not a unidirectional process; reverse mutations occur. A more realistic picture is thus

$$a^{+} = \Leftrightarrow a$$

in which the "wild-type" allele a+ "wild-type" gene to a mutant form (usually a nonfunctional or only slightly functional form) is investigated, what is actually measured is the summation of many different mutation events at many different sites in the gene. The "average gene" is usually considered to be a DNA sequence of about 1000 nucleotide pairs coding for polypeptide that is about 333 amino acids long. Amino acid changes at many different positions in the polypeptide (resulting from base-pair substitutions at many different positions in the structural gene) may be expected to result in an inactive gene product. Thus, when one measures the mutation rate of a "wild-type" gene to the nonfunctional mutant from, one will actually be measuring the sum of many different mutations, each occurring at its own specific rate.

Mutation "Load" Versus Genome Size

In the preceding section, an average mutation rate of 1 per 100,000 (or 10^{-5}) per gene per generation was cited for Drosophila. Estimates of the total mutation rate–the summation of mutation rates of the entire genome - in Drosophila are of the order of 5 percent (or 5 × 10^{-2}) per generation (per haploid complement, or per gamete). These values can be used to estimate the number of genes in the Drosophila genome (haploid complement). If the mutation rate per gene in m and the mutation rate per total genome is M, then the number of genes in the genome (N), will equal M/m. Using the above estimates, N = 5 × 10–2/10–5 = 5000. This estimate agrees well with estimates of the number of genes in Drosophila obtained by other genetic criteria. On the other hand, one biochemical estimate suggests that Drosophila has about three times that many genes.

In the Drosophila genome (haploid) contains only 5000–15,000 genes, there is a major difference in the organization of the Drosophila genome and the genomes of the extensively studied bacteria and viruses. This difference, in fact, appears to be a basic difference between prokaryotes and eukaryotes. In prokaryotes, essentially all of the DNA of the genome appears to represent structural genes. Estimates of the number of genes from mutation studies (in some case, identification of all or most of the genes of the organs) agree well with estimates obtained by dividing the total DNA content in nucleotide pairs) of the genome by 1000 (103 nucleotide pairs per "average" gene). The phage lambda chromosome consists of 45 × 10^3 nucleotide pairs and about 40 known genes. The phage T4 chromosome contains about 200 × 10^3 nucleotide pairs. Over 100 genes have been mapped in phage T4, and a significant number remain to be identified. The E. coli genome contains about 4000 × 10^3 nucleotide pairs, while mutation analyses suggest that it contains about 3000 genes. In these prokaryotic organisms, then, the genomes are composed almost entirely of structural genes. The Drosophila melanogaster haploid chromosome complement contains about 120,000 × 10^3 nucleotide pairs. If all of the DNA represented structural genes of average size (103 nucleotide pairs), then Drosophila should have about 120,000 genes, rather than 5000 to 15,000 genes as indicated by the calculation above.

It now seems clear the most of the DNA of Drosohila and other eukaryotes, including humans, does not represent structural genes. What is the function of this "excess" or noncoding DNA? Many geneticists believe that this noncoding DNA is regulatory, playing important roles in the regulaion of gene expression. Others believe that it has an important role in some aspect of chromosome structure. Still others believe that much of the noncoding DNA is just "junk" DNA with no important function, possibly a reservoir of nucleotide-pair sequences available for the evolution of new genes. In any case, this DNA does not code for proteins. This "noncoding" DNA is of two types: DNA that is transcribed, but for which the transcripts never leave the nuclei, and (2) nontranscribed DNA. These two types of DNA may have very different functions. The first type of noncoding DNA clearly includes many of the noncoding intervening sequences or introns of eukaryotic genes.

An estimate of the maximum number of essential genes per genome can also be made from considerations of average mutation rates and the maximum mutation "loads" that a species could be expected to survive. For example, consider an average mutation rate of 4×10^{-6} recessive lethal mutation per essential gene per generation (per gamete). J. L. King, and others, have emphasized that it is unlikely that a species could tolerate more than 0.8 new lethal mutation per zygote (or 0.4 per gamete). Dividing 0.4 new recessive lethal mutation per gamete by 4×10^{-6} new recessive lethal mutation per essential gene per gamete gives a maximum number of 10^5 essential genes per gamete. At 10^3 nucleotide pairs per gene, the 105 essential genes would represent a total for 10^8 nucleotide pairs. The haploid chromosome complement of mammals, including humans, contains about 3×10^8 nucleotide pairs. Thus, according to thee estimates, essential genes can represent a maximum of a only about 3 percent of the total DNA in the genome of mammals. The actual number of essential genes in mammals is probably closer to $1\text{–}2 \times 10^4$, or less than 1 percent of the total DNA.

Mutable Genes

Most gene are relatively stable and mutate infrequently, but in some organisms a few genes mutate spontaneously so often that individuals carrying them are mosaics of mutated and unmutated genes. The R locus, which is involved with anthocyanin pigment synthesis in maize, for example, was found by R.A. Emerson to undergo alteration much more frequently than other loci. The change from Rr to rr, for example, was found to occur, at the rate of 1 per 20,000 gametes. These "mutable" genes are either more unstable than others or they are influenced by other factors in the genetic environment. McClintock's conclusion from extensive studies on maize was that a mutable gene is not an autonomous entity, but is derived from an agent that is integrated at the site of the mutable gene to cause instability.

Examples of highly mutable genes have been found in both plants and animals, but they appear to be more common in plants. Mutations in highly mutable genes occur frequently in somatic tissue and occasionally in germ cells. Somatic mutations may show their effects as colour variegations (mosaics) in such plant part as endosperm, leaves, and petals. Many common plants—including the larkspur, snapdragon, sweet pea, four-o'clock, and morning glory—have colour variegations suggesting unstable or mutable genes.

The classical investigations of M. Demerec (1941) on a mutable gene called miniature-alpha in Drosophila virilis provided the first substantial data on the genetic properties of mutable gens in animals. These propertie are: (1) mutation occurs primarily before meiosis; (2) mutation occurs in both females and males, implying that meiotic crossing over is not involved (crossing over occurs only rearely in Drosophila males) and (3) mutation is strongly influenced by neighbouring genes. More recently, M.M. Green and others have described several mutable gene systems involving the white locus (eye colour) and other genes in D. melanogaster. White-crimson (w^c), for example, mutates to wild-type and phenotypes other than white-crimosn at a frequency of 10–3. This mutable system and several others at the w locus are associated with chromosome alterations, particularly deficiencies. This led to the hypothesis that a "*controlling element*" from the cytoplasm is integrated into the chromosome at the site of mutability, and that this agent is responsible for chromosome aberrations.

Many of these "controlling elements" have now been identified by G. Rubin, P. M. Bingham, and colleagues, and shown to be transposable elements with structures similar to the TN elements of bacteria. The unstable allele ww^c mentioned above has been shown to result from the insertion of an approximately 10,000 nucleotide-pair-long transposable element belonging to the class called FB (for "foldback") elements.

Another unstable allele, w^{a1} has been shown to have a transposable element called *copia* (because its sequence is present in RNA in *copious* amounts) inserted at the site of mutation. Copia has been isolated and sequenced. It is about 5,000 nucleotide pairs long with 276 nucleotide-pair perfect direct repeats at its ends. These 276 nucleotide-pair repeats, in turn, have 17 nucleotide-pair inverse repeats at each end. Note the similarity of *copia* to E. coli elements Tn 9 and Tn 10.

Mutator and Antimutator Genes

Genes in maize were shown by McClintock to influence the stability of other genes. These have been called mutator genes. A striking example of the action of a mutator gene with a specific effect on a basic colour gene in maize was described by M. M. Rhoades. The colour of maize leaves and other plant parts is dependent on a complex of three complementary genes, symbolized A, C and R. A colour other than green is produced only when the dominant alleles (A, C, and R) of all three genes are present. The colour may be purple if gene P is also present, or red (pp), or some other colour depending on what other genes are included along with A-C-R. Plant soft genotype aa, cc, or rr have green leaves regardless of alleles present at the other loci. Plants with the genotype aaC-R- would be expected to be green, but in the presence of a mutator gene called Dt, they are variegated. Light-coloured corn kernels have purple spots at locations where somatic mutations have occurred. The Dt gene produces its effect by influencing one or more of the a alleles of the aa genotype to mutate to A. Patches of cells scattered throughout the plant carry A alleles resulting from these mutate to A. Patches of cells scattered throughout the plant carry A alleles resulting from the mutations. On the leaves and kernels, these patches give a speckled appearance. The size of the spots depends on the stage of development at which the mutations occurred. Green cells contain unmutated genes; that is, they are of genotype aa. In this case, the allele (Dt or dt) present at one locus thus influences the mutation rate of an allele (a) present at another locus.

J. F. Speyer and others have found broad-spectrum mutator genes in bacteriophage T4. Temperature sensitive (ts) mutants of gene 43, for example, alter the mutation rates of point mutations in other genes. Some mutator genes exert their mutagenic effect during DNA replication by altering polymerase activity, producing mutagenic base analogs, or modifying DNA bases, thus influencing the mutation rates of other genes. One mutator in phage T4 gene 43 (ts L88) produces a DNA polymerase that utilizes incorrect nucleotides at a higher frequency than does the wild-type enzyme.

Several mutations in gene 43 exhibit powerful negative or antimutator activities, particularly during the formation of A : T → C : G substitutions. Experiments have shown that DNA polymerase (gene 43 products) that are isolated from E. coli that were infected with mutator, antimutator and wild-type strains for T4 bacteriophage discriminate between adenine and 2-aminopurine to different degrees during DNA synthesis in vitro. Significantly larger amounts of 2-aminopurine are incorporated into DNA by wild-type and mutator than by antimutator enzymes. This indicates that organisms have the potential to evolve mechanism that result in mutation rates that are lower than those presently existing in wild-type strains of somatic organisms.

When the mutator and antimutator DNA polyemerase of phage T4 were studied in vitro, they were found to have altered ratios of polymerase activity to 3'→5' exonuclease activity. Mutator have increased polymerase/exonuclease activity. Antimutators exhibit decreased polymerase/exonuclease activity. This suggests that the mutation rate is, at least in part, controlled by the relative rates of polymerization and proofreading rates decrease mutation rates (antimutator). Decreased proofreading efficiencies (or increased polymerization rates) increase mutation frequencies (mutator).

Several mutator genes have also been identified and characterized in E. coli.Some increase the frequency of only certain types of base-pair substitutions. Others cause an increase in all types of point mutations. R. C. von Borstell and others measured spontaneous mutation rates in yeast and isolated many mutator genes with different kinds of mutator and antimutator activity.

Decreases as well as increases in mutation rates, as compared with wild-type, can thus result from new mutations. With both mutators and antimutaors operating in the same system, particular spontaneous mutation rates may be optimized through natural selection. Mutators and antimutators thus become relative terms, because no standards are available for optimum rates. Results of studies on yeast have borne out the conclusions of those on viruses and bacteria; that spontaneous mutation rates are under the genetic control of the cell itself.

Insertion Sequence ("Is Elements") As Mutators

A significant proportion of the mutations that occur in bacteria and some bacteriophages are now known to be effects of small sequences of DNA called IS elements or insertion sequences. These DNA sequences, from about 800 to about 1400 nucleotide paris in length are present in E. coli chromosomes, for example, in several copies. The first five IS elements to be characterized were: ISI (768 nucleotide pairs long) and IS2, IS3, IS4, and IS5 (all between 1200 and 1400 nucleotide pairs in length). The IS elements are transposable; that is, they have the ability to move from one position in the chromosome or to other chromosomes in the same cell. They can move, for example, from the E. coli chromosome to bacteriophage chromosomes or to plasmids (small circular molecules of DNA, or "minichromosomes," often present in bacteria.

The chromosome of the strains of E. coli that have been studied contain about eight copies of ISI elements are physically and covalently inserted as linear sequences into the chromosomes when an IS element is inserted (or insert itself?) into a gene, it destroys the function of, mutates, that gene. While insertion of IS elements occur at many sites in the E. coli chromosomes, it is not completely at random. Preferred sites of insertion have been demonstrated in some case.

Detectable IS-induced mutations occur with a frequency of about 10^{-6} to 10^{-7} per cell generation in E. coli. IS-induced mutations are revertible. They can be distinguished from all mutations with similar effects in that their reversion frequency (usually in the range 10^{-6} to 10^{-7}) can not be enhanced with mutagens. The fact that IS-induced mutations are revertible means that IS elements can excise themselves from the chromosome with precision to the single nucleotide pair. Reversion to the functional state requires restoration of the original nuclotide-pair sequence of the gene; no base-pairs can be added or deleted during the insertion-excision process. The failure of mutagens to hence reversion is to be recombination mechanism, not a mutation process.

The "controlling elements" and mutable loci (or unstable genes) in higher plants and animals are probably caused by sequences of DNA analogous to the IS elements in prokaryotes.

Induced Mutations

Mutations can be induced by either physical or chemical means: such mutations are called induced mutations. Irradiation is an example of a physical mutagen, with x-rays, gamma-rays, and ultraviolet light being the most common examples used. One consequence of x- or gamma-ray irradiation is the breakage of chromosomes, which may result in chromosomal rearrangements, or the events may be lethal to the cell.

5-Bromouracil

5-Bromouracil (5-BU) is a base analog; that is, its structure closely resembles one of the bases normally found in DNA. 5-BU can exist in two states. In its usual keto state it exhibits properties similar to those of thymine and thus will pair with adenine in DNA. Rarely it switches to the enol state, and in this form it will pair specifically with guanine.

Mutations can be induced by 5-BU (and in general by base-analong muta-gens) in two ways. The first involves the incorporation of the usual form of 5-BU into DNA during replication. If 5-BU shifts to its rare enol state during the next round of replication, the result will be a transition mutation from AT to GC.

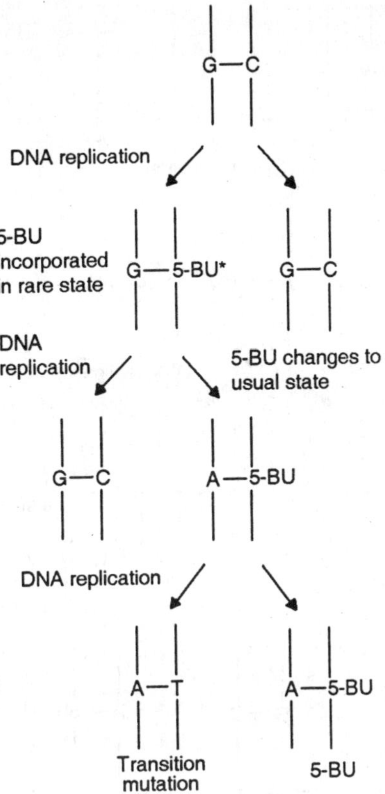

Fig. 10.7. Mutagenic action of 5-BU when it incorporates into DNA in its usual keto state and then shifts to its rare state during the next round of replication.

The second way that 5-BU can induce mutations is if the base analog g is incorporated in the DNA while it is the rare enol state. This dictates insertion opposite a G on the complementary strand, and subsequent replication with a shift of the 5-BU to the usual keto state will result in a transition mutation from GC to AT.

Thus 5-BU can induce either AT to GC or CG-to-AT transition mutations. (In the jargon of this area, 5-BU is said to induce two-way transition mutations.) Therefore it is possible to correct a 5-BU induced transition mutation by treating with 5-BU for a second time. This is called *reversion* of the mutation.

2-Aminopurine

2-Aminopurine (2-AP) is also a base analog and, like 5-BU, it can exist in two states. In its usual state it behaves like adenine and will

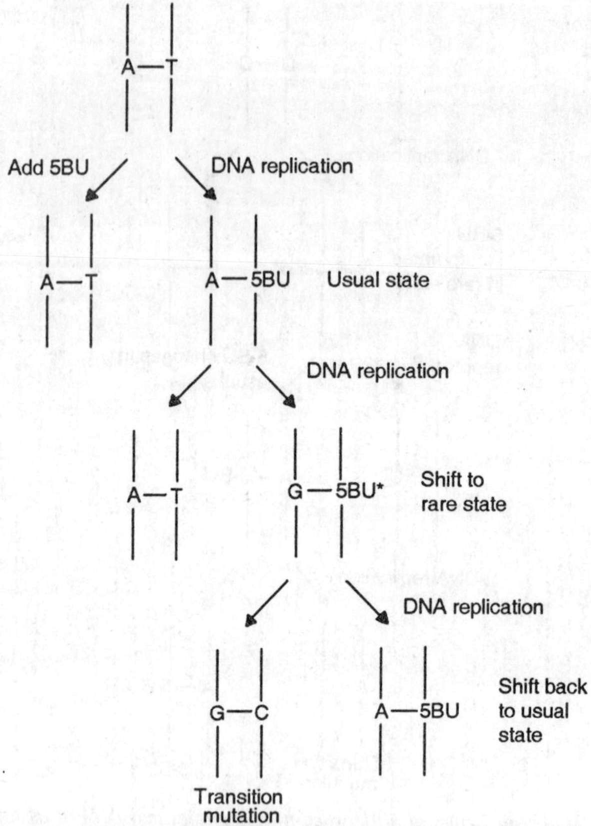

Fig. 10.8. Mutagenic action of 5-BU when it incorporates into DNA in the rate state and then shifts to the usual keto state during the next round of replication.

form two hydrogen bonds with thymine. In its rare amino state 2-AP behaves like guanine and forms two hydrogen bonds with cytosine. Thus 2-AP can induce transition mutations both from AT-to-GC and from GC-to-AT. 2-AP induced mutations can therefore be reverted by 2-AP treatment.

Nitrous Acid

Nitrous acid (NA : HNO_2) is a deminating, agent; it acts by removing amino groups (NH_2) from the bases. In some but not all instances this alters their base-pairing abilities and hence induces mutations. The three bases that have amino groups are adenine, guanine, and cytosine. When adenine is treated with NA, it is changed to hypoxanthine, which will pair with cytosine. This results in an AT-to-a

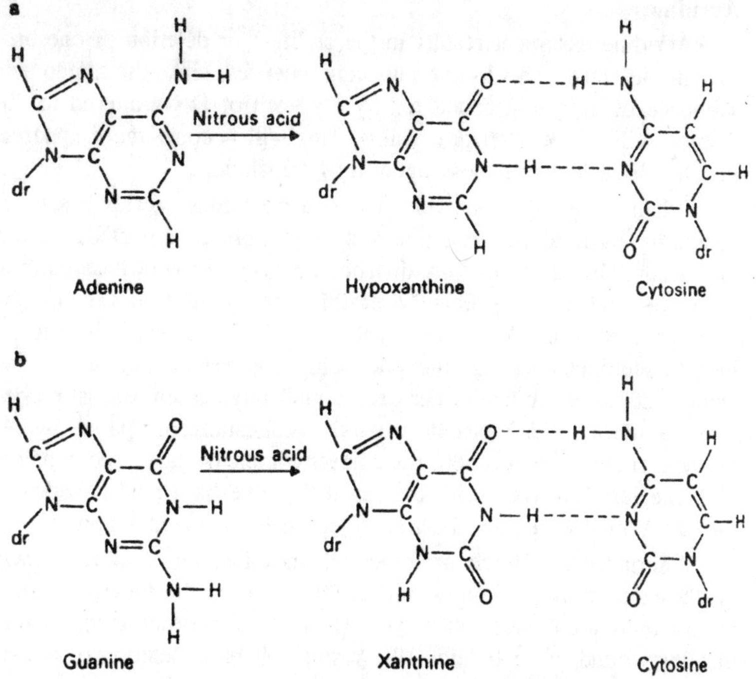

Adenine → Nitrous acid → Hypoxanthine · Cytosine

b

Guanine → Nitrous acid → Xanthine · Cytosine

Fig. 10.9. Mutagenic action of nitrous acid. (a) Deamination of adenine by nitrous acid treatment produces hypoxanthine, which pairs with cytosine, (b) Deamination of guanine produces xanthine, which pairs with cytosine.

GC transition mutation. Treatment of guanine with NA removes the amino group from the 2-carbon position and produces xanthine. However, since both guanine and xanthine pair with cytosine, no base-pair mutation results.

Deamination of cytosine by NA produces uracil, which, or course, pairs with adenine. This results in a GC-to-AT transition mutation–the opposite of NA's effect on adenine. Thus mutations induced by nitrous acid can be reverted by nitrous acid treatment: in other words, nitrous acid induces two-way transition mutations.

Hydroxylamine

Hydroxylamine (NH_2OH) reacts only with cytosine, hydroxylating it so that it can then only with adenine . Thus hydroxlamine induces one-way transition mutation from GC to AT. Because of this, mutations induced by hydroxylamine cannot be reverted to revert by 5-BU. 2-AP, or NA treatment since these mutagens can bring about a GC-to-AT transition.

Acridines

Acridine treatment results in the addition or deletion of one base pair in the DNA. This has serious consequences, since the amino acid sequence of a protein coded for by a stretch of DNA altered in this fashion will be changed drastically. This will become more apparent in later discussions of messenger RNA translation.

When present at relatively low concentrations, acridine acts by becoming inserted between adjacent base pairs in the DNA. When this occurs, it "stretches" the distance between adjacent base pairs to 0.68 nm, which is precisely double the normal distance. The consequences of this depend on whether the acridine molecule is inserted into the template strand (the one being copied) or into the strand being synthesized. In the former case, a randomly chosen base is inserted opposite the acridine molecule when the replication fork passes by. At the next round of replication, the correct complementary base is paired with the inserted base, with the result that one base pair is added to the DNA in that region. This is called an insertion mutation.

Alternatively, if the acridine becomes inserted into the newly synthesized strand, it blocks one of the bases on the template strand from having a complementary base Then, if the acridine is lost before the next round of replication, the result will be a deletion of a base pair. Therefore it is possible to revert an acridine-induced mutation by a second treatment with acridine.

In summary, the mutagens described in this section have different modes of action and cause different mutation changes.

Table 10.2. Summary of the method of action of various chemical mutagens.

Mutagen	*Base-pair changes*
5-Bromouracil	AT ↔ GC two-way transitions
2-Aminopurine	AT ↔ GC two-way transitions
Nitrous acid	AT ↔ GC two-way transitions
Hydroxylamine	GC → AT one-way transitions
Acridines	+1 or −1 insertion or deletion

Some of the chemical mutagens described in this section are commonly used in the laboratory. Many other chemicals appear to cause mutations, and public awareness of this is increasing as industrial effluents, cosmetic ingredients, food additives, etc. are examined carefully for any mutagenic activity in test organisms.

11

Human Genome

In evolution, novel genes have arisen either by whole genome duplication or by regionally localized duplication events. Both mechanisms give rise to *paralogous* genes, genes that occur within the same species and which have a common ancestor. Thus, members of multigene families and superfamilies are paralogous and are distinct from *orthologous* genes which are found in different species and have diverged from their common ancestor over evolutionary time. In practice, the consequences of whole genome duplication and regionally localized gene duplication are often very hard to distinguish.

Ohno (1970) first proposed that the vertebrate genome had evolved to its present size through an ancient tetraploidization event. Evidence supporting this hypothesis now comes from a variety of different sources. Comparative nuclear DNA measurements in higher organisms are compatible with the idea of successive genome doublings. So are cytogenetic data, since the human karyotype can be divided into similar pairs of chromosomes on the basis of structure and banding patterns viz. chromosomes 1 and 2, 4 and 5, 7 and 8, 11 and 12, 14 and 15, 16 and 17, 19 and 20, 21 and 22. The postulated ancient tetraploidization event also appears to be reflected in the genetic information content of these chromosome pairs. This is exemplified by the distribution of the insulin and RAS oncogene family members in the human genome: the insulin (INS), insulin-like growth factor 2 (IGF2) and Harvey RAS (HRAS) genes are located on chromosome 11p whilst the insulin-like growth factor 1 (IGF1) and Kirsten RAS (KRAS2) genes are located on chromosome 12.

That the genome duplications occurred prior to the adaptive radiation of the vertebrates is evidenced by the similar gene number exhibited by both fish and mammals, the survival of syntenic linkage groups over quite long periods of evolutionary time and the tetralogy exhibited by many vertebrate loci. This tetralogy is evident from the observation that the basic set of ~15,000 genes found in all primitive metazoans varies only moderately from *Caenorhabditis elegans* to *Drosophila* to *Ciona intestinalis*, a tunicate but this gene complement is approximately four-fold smaller than the total number of genes in vertebrate genomes. At the level of the individual gene, there are numerous instances where an invertebrate (*Drosophila*) gene has up to four related genes (paralogues) in vertebrates, implying two rounds of genome duplication. These paralogues represent quadruplicated loci on different chromosomes that are more similar to each other than they are to members of other tetralogous groups. In humans, these so-called 'tetralogues' include the HOX genes (HOXA, 7p14-p15; HOXB, 17q21-q22; HOXC, 12q12-q13; HOXD, 2q31), the epidermal growth factor receptor genes (EGFR, 7p13; ERBB2, 17q11.2-q12; ERBB3, 12q13; ERBB4, 2q34), the Jak family of tyrosine kinases (jAK1, 1p31.3-32.3; jAK2, 9p24; jAK3, 19p12-p13; TYK2, 19p13.2), the MADS box enhancing factors (MEF2A, 15q26; MEF2B, 19p12; MEF2C, 5q14; MEF2D, 1q12-q23), the Src family of nonreceptor tyrosine kinases (SRC, 20q11.2; YES1, 18p11.22-p11.31; FGR, 1p36.1-p36.2; FYN, 6q21) and the syndecans (SDC1, 2p; SDC2, 8q22-q23; SDC3, 1p32; SDC4, 20q12-q13). Many further examples of triplicated loci occur in the human genome. In these cases, it may be that one paralogue has been lost or alternatively, still remains to be characterized.

The results of studies of the genes encoding the homeobox proteins, insulin-like growth factor genes and high mobility group proteins in the primitive chordates *Amphioxus* and *Ciona*, and a jawless vertebrates *Lampetra fluviatilis* (lamprey), are consistent with the occurrence of one genome duplication in the common ancestor of all jawed and jawless vertebrates after the lineage leading to *Amphioxus* had diverted and a second genome duplication occurring in the common ancestor of jawed vertebrates after their divergence from the jawless vertebrates. Studies of HOX gene number suggest that the duplication events are likely to have occurred before the radiation of the teleosts.

Gene number does not however automatically distinguish between tandem duplications and polyploidization events. Postlethwait et al. (1998) mapped 144 zebrafish (*Danio rerio*) genes; comparison of the resulting map with their mammalian counterparts led to the identifications of orthologous chromosome segments for at least three chromosome paralogy groups in zebrafish and mammals. This finding is consistent with the hypothesis that these segments were duplicated prior to the divergence of zebrafish and mammals. The presence of more than two copies of each paralogous chromosomal segment, is suggestive of at least two rounds of duplication which would have occurred after the divergence of the cephalochordates and cranial chordates, but before the divergence of the ray-finned and lobe-finned fishes, which is thought to have occurred about 420 Myrs ago.

Table 11.1. Possible paralogies between parts of human chromosomes 4 and 5

4	5
FGFR3	FGFR4
	HTR1A
	ADRB2
ADRA2C	ADRA1B
DRD5	DRD1
QDPR	DHFR
GABRA2	GABRA1
GABRB1	
STATH	SPARC
KIT	PDGFRB
PDGFRA	CSF1R
AREG	C7
EGF	C9
AGA	HEXB
FGF5	
FGF2	FGF1
IF	F12
F11	GZMA
KLK3	
	CSF2
IL2	IL3
	IL4
	IL5
	IL9
	CSF1
MLR	GRL
ANX3	
ANX5	ANX6

Several extensive regions of paralogy have been identified in the human genome which have been claimed to result from ancient tetraploidization events. The 13 groups of paralogous genes found on chromosomes 4 and 5 provide one example. Lundin (1993) identified several other possible examples of paralogous pairs or groups of genes on different human chromosomes; (i) parts of chromosomes 2, 7, 12, 14, and 17, (ii) parts of chromosomes 8, 10, and 16, and (iii) parts of chromosomes, 1, 11, 12, 15, and 19. Although the extensive paralogy noted between chromosomes 11 and 12 is explicable by a model of chromosome duplication resulting from tetraploidization, there are some discrepancies in the location of genes on these chromosomes. These can however be accounted for by the occurrence of a pericentric inversion on chromosome 12.

Paralogy may be explained by mechanisms other than tetraploidization. Indeed, some paralogous gene loci are explicable by regional duplication. The relative importance of regional duplication/translocation as compared to tetraploidization is unclear and ambiguity even extends to individual cases. Thus, the relative locations of the tyrosine hydroxylase (TH; 11p15.5), tryptophan hydroxylase (TPH; 11p14.3-p15.1) and phenylalanine hydroxylase (PAH; 12q22-q24) genes have been explained in terms of both mechanisms.

The two highly related regions on the proximal and distal long arms of human chromosome 21 (21q22.1 and 21q11.2) appear to have arisen as a result of an intrachromosomal duplication of >200 kb. This duplication is thought to have arisen between 15 and 30 Myrs ago after the separation of the orangutan from the other great apes. By contrast, the origin of the paralogous 2-20 Mb segments on human chromosomes, 1, 6, and 9 is unclear. Regardless of the mechanism, at least two intra-chromosomal duplications must have occurred resulting in the triplication of a series of genes, for example the retinoid X receptor genes and RXRG (1q22-q23) RXRB (6p21.3) and RXRA (9q34.31), the pre-B cell leukemia transcription factor genes PBX1 (1q23), PBX2 (6p21.3), and PBX3 (9q34), and the tenascin genes TNR (1q25-q31), TNXA (6p21.3), and HXB (9q32-q34). Interestingly, *Alu*- and LINE-dense clusters flank the boundaries of the 6p21.3 segment, a finding which may be significant in view of the recombinogenic potential of these sequence elements. In this context, it may be significant that a sequence related to the pseudoautosomal boundary of the human sex chromosomes has also been noted at the centromeric boundary of the 6p21.3 segment.

Table 11.2. Possible paralogies between parts of human chromosomes 1, 11, 12, 15 and 19

1	11	12	15	19
RYR2		CACNA1C		RYR1
		BCAT1		BCAT2
EN01		EN02		
GDH			SORD	
EBVS1	EBVM1			
				CEAL1
				CEA
PTPRF	NCAM1			MAG
				NCA
				PSG1-13
TNFR2		TNFR1		
SLC2A1		SLC2A3		
SLC2A5				
GOT2L1		GOT2L3		
GOT2L2				
GNAI3		GNA12L		
GNAT2				
GNB1		GNB3		
	LMO1	LMO3		
	LMO2			
MYCL1		MYF5		LYL1
MYOG	MYOD1	MYF6		TCF3
CHRM3	CHRM1,4		CHRM5	
	FGF3	FGF6		
	FGF4			
		A2M		
		PZP		
	ESA4	ELA1		C3
	F2	C1S		KLK1
		C1R		KLK2
			LIPC	LIPE
		TP11	MPI	GPI
LDHAL2	LDHA, LDHC	LDHB		
NRAS	HRAS	KRAS2		
RAP1A		RAP1B		RRAS
RAB3B				RAB3A

1	11	12	15	19
RAB4				
	CALCA	IAPP		
	CALCB			
	PTH	PTHLH		
FGR				
LCK	SEA		FES	
ABL2				
JUN				JUNB
				JUND
INSRR		ERBB3	IGF1R	INSR
NTRK1			LTK	TYK2
C8A, C8B		LRP1	THBS1	LDLR
GSTM1	GSTP1	MGST1		
C1QA, C1QB				
COL11A1		COL2A1		
COL8A2				
H1F2		H1F4		
	PGR	HMR		
		RARG	CRABP1	
		VDR		
HKR3	WT1	GLI		HKR1
				HKR2
PEPC		PEPB	ANPEP	PEPD
KCNC4		KCNA1,2,5		KCNA7
		KCNC2		KCNC3
ACADM		ACADS	IVD	
	TH, TPH	PAH		
NGFB	INS			
	IGF2	IGF1		
	INSL2			
	TRV2	TRV3		
	FRV1	FRV3		
PLA2G2A		PLA2G1B		
PLA2G2C				
TSHB	FSHB			LHB
				CGB
	ACP2		ACP5	
PKLR			PKM2	
ATP1A1				ATP1A3
ATP1A2				

1	11	12	15	19
ATP1B1				
ATP1AL2		ATP2A2		
ATP2B2		ATP2B1		
TRAP2		TRA1	TRAP1	
CD48	THY1			
	CD3D			A1BG
	CD3E			
CD3Z	CD3G			
CD1AE	FCER1B			
FCER1A				FCER2
FCER1G				
FCGR2A, 2B				
2C, 3A, 3B		CD4		
PIGR				
	APOA4			APOC1
APOA2	APOA1			APOC2
	APOC3			APOE
	FUT4			FUT1
	SIAT4C			FUT2
			MANA1	MANB
			CKMT1	CKM
	GANAB		GANC	
CAPN2	CAPN1			
			CAPN3	CAPN4
MUC1	MUC2,			
	MUC5B			
	MUC5AC			
AT3	C1NH			
AGT				
PFKM		PFKX		
FDPSL1			CHR39B	
ACTA1			ACTC	
POU2F1				POU2F2
TNNI1				TNNT1
SNRPE			SNPRN	SNRPA
				SNRP70
TGFB2				AMH
				TGFB1
CTSE	CTSD		CTSH	
REN	PGA3-5			

Table 11.3. Locations of human gene loci indicating large regional duplications of chromosome 1

TRN	tRNA, asparagine	1p36
TRN	tRNA, asparagine-like	1q12-q22
TRE	tRNA, glutamic acid	1p36
TREL1	tRNA, glutamic acid-like 1	1q21-q22
RNU1	Small nuclear U1 RNA	1p36
RNU1P1-4	Small nuclear U1 RNA pseudogenes 1-4	1q12-q22
FGR	Feline sarcoma oncogene	1p36
LCK	Lymphocyte protein tyrosine kinase	1p32-p35
ABLL	Abelson murine lukemia oncogene-like	1q24-q25
C8A, C8B	Complement component, 8, α/β chains	1p22-p36
C1QA, C1QB	Complement component 1q, α/β chains	1p
C4BPA, C4BPB	Complement component 4 binding protein	1q32
CR1, CR2	Complement component receptor 1/2	1q32
AK2	Adenylate kinase 2	1p34
GUK1	Guanylate kinase 1	1q32-q42
GOT2L1	Glutamic-oxaloacetic transaminase 2-like 1	1p32-p33
GOT2L2	Glutamic-oxaloacetic transaminase 2-like 2	1q25-q31
RAB3B	Member RAS oncogene family	1p31-p32
NRAS	Neuroblastoma RAS oncogene	1p13
RAP1A	Member RAS oncogene family	1p12-p13
RAB4	Member RAS oncogene family	1q42-q43
FTHL1	Ferritin, heavy polypeptide-like 1	1p22-p31
FTHL2	Ferritin, heavy polypeptide-like 2	1q32-q42
CD58	Lymphocyte function-associated antigen	1p13
DAF	Decay accelerating factor for complement	1q32
ATP1A1	ATPase, Na^+ K^+, $\alpha 1$ polypeptide	1p13
ATP1A2	ATPase, Na^+ K^+, $\alpha 2$ polypeptide	1q21-q23
ATP1B1	ATPase, Na^+ K^+, β polypeptide	1q22-q25

Consequences of Genome Duplications for Gene Evolution

In principle, the genetic redundancy created by a genome duplication would have allowed evolutionary experimentation, in that while one gene copy continued to function as before, the other was freed to acquire mutations, irrespective of whether they were adaptive or inactivating. If the newly duplicated gene acquired mutations that modified either the expression pattern of the encoded gene or the function of the encoded protein in an advantageous way, the novel allele could have become fixed in the population. Ohno (1970) expressed this idea rather elegantly: 'An escape from the ruthless pressure of natural selection is provided by the mechanism of gene duplication. By duplication, a redundant copy of a locus is created. Natural selection often ignores such a redundant copy, and, while being ignored, it accumulates formerly forbidden mutations and is reborn as a new gene locus with a hitherto non-existent function. Thus, gene duplication emerges as the major force of evolution.'

In this context, evidence for positive selection has come from the observation of accelerated evolution in some genes subsequent to gene duplication. In such studies, positive selection is implicated in the process of evolutionary change by the observation of a higher frequency of non-synonymous over synonymous substitutions. Changes in the expression patterns of the duplicated genes are sometimes also apparent as in the neuronal and muscle expressed genes of the nicotinic acetylcholine receptors family. In principle, gene conversion may either promote diversification of proteins encoded by duplicated genes or promote homogenization in which case the molecular evolutionary record is automatically erased. In practice, however, at least for the HLA and immunoglobulin genes, gene conversion is notable more by its absence: inserted new genes tend to be created by a 'birth-and-death' process of duplication and deletion.

A more likely scenario, however, is that the duplicated gene rapidly acquires inactivating mutations and becomes a pseudogene. Indeed, assuming that a gene duplication is not selectively disadvantageous, the duplicated genes can survive in the genome for quite long periods. Arguably the best available model system to assess whether duplicated genes will be retained or inactivated over evolutionary time is yeast (*Saccharomyces cerevisiae*), the organism with the best characterized genome duplication. Sequencing data from the 12 Mb yeast genome are consistent with the occurrence of a whole genome duplication that occurred ~100 Myrs ago, after the divergence of *S. cerevisiae* from *Kluyveromyces*. Most duplicated genes were subsequently deleted; only 13% of yeast proteins are now represented as homologous pairs encoded by homologous genes. Of these homologous gene pairs, only a few possess functions that have clearly diverged under the influence of selection viz. the mitochondrial and peroxisomal isozymes of citrate synthase (CIT1 and CIT2), RAS1 and RAS2, the transcription factors ACE2 and SW15, the phosphatidylinositol kinase TOR1 and TOR2, and the myosins MYO3 and MYO5. Conclusions drawn from yeast may not however be applicable to vertebrates and in this phylum, many more genes may have survived the aftermath of duplication to acquire new functions. Thus, Nadeau and Sankoff (1997) analyzed the frequency of distribution of family size for gene families present in humans and mice and which arose putatively by genome duplication early in vertebrate evolution. They concluded that duplicated genes were as likely to have survived and acquired a novel function as to have been lost through the acquisition of inactivating mutations. In agreement with these findings, studies of fish and *Xenopus* have both suggested that about 50% of newly duplicated genes are retained after tetraploidization.

Mammalian Genome Evolution

The best estimates of divergence times for the various mammalian orders and the other major vertebrate lineages have come from the use of extant gene sequences to calibrate a 'molecular clock' of vertebrate evolution. Gene-specific evolutionary rates often vary quite dramatically, but the use of multiple genes to derive mean divergence times should yield more accurate and reliable estimates. Kumar and Hedges (1998) therefore employed 658 genes from 207 vertebrates specie to derive a molecular timescale for vertebrate evolution. The divergence times corresponded well to previous estimates based upon the fossil record. Thus the calculated divergence time for the jawless fish (Agnatha) was 564 Myrs ago in the Precambrian era. Interestingly, the molecular data indicated that at least five major lineages of placental mammals [Edentata (armadillos, anteaters and sloths), Hystricognathi (porcupines and guinea pigs), Sciurognathi (squirrels), Paenungulata (hyraxes) and Ferungulata (carnivores)] could have arisen in the early to middle Cretacious between 130 and 90 Myrs ago. This represent an important revision of previous estimates of the timing of the adaptive radiation of the mammals. Since this now appears to have predated the Cretacious/Tertiary extinction of the dinosaurs 65 Myrs ago, the adaptive radiation of the mammals could not have been simply a consequence of the filling of niches vacated by the departing super-lizards. Other factors such as climatic changes and the continental breakup must also have played a role. This question notwithstanding, the adaptive radiation of the mammals has been very successful, resulting in the emergence of >4600 living species that occupy a very diverse range of habitats and environments.

It has been suggested that the rate of mammalian speciation may have been influenced by the rate of karyotypic change. However, mammalian genomes still contain significant regions of genetic linkage

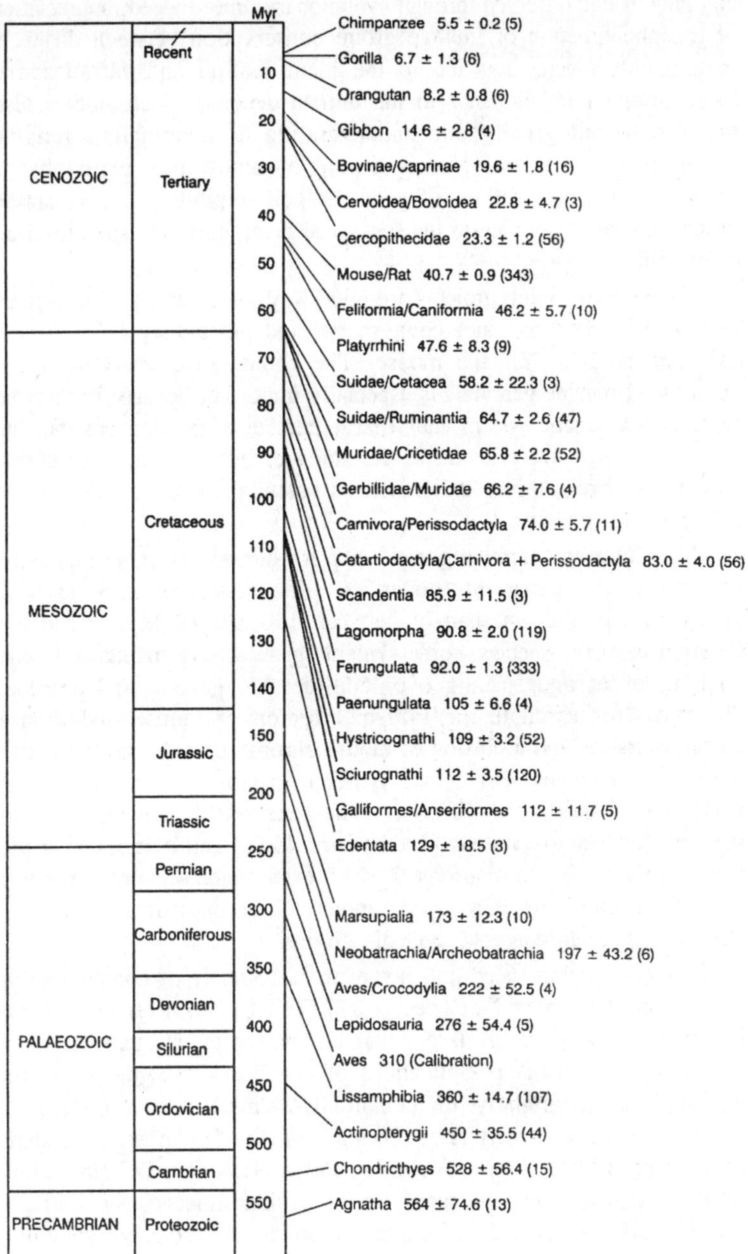

Fig. 11.1. A molecular timescale for vertebrate evolution.

that have been conserved through evolutionary time. Indeed, appreciation of the phenomenon of linkage group conservation between different mammalian species has led to the identification and chromosomal localization of novel genes in the human genome. Such studies also promise to aid greatly our understanding of mammalian genome evolution by, for example, revealing chromosomal inversions or translocations, duplications of genes or gene regions, or more subtle lesions that may have led to the functional divergence and specialization of proteins.

Genomic mapping projects are underway for a variety of mammals including the mouse, rate, cow, sheep, and pig but by far the most data are available for the mouse. The comparative analysis of the human and murine genomes is especially important because the mouse represents a genetic system with a large number of spontaneous mutants (some of which are relevant to the study of human genetic disease), numerous inbred strains and enormous potential for the application of transgenic technology.

The latest comparative genetic map for human and mouse contains nearly 1800 genes mapped to over 200 different syntenic chromosomal groups. Despite a timespan of between 100 and 80 Myrs since the divergence of the species, some syntenic groups have maintained gene content, order and spacing over considerable genetic and physical distances, for example the lq21-q23 region of human and mouse chromosomes 1 and a portion of mouse chromosome 2 with the entire human chromosome 20. Other syntenic groups, by contrast, have accumulated substantial differences in gene order between the two species, for example human chromosome 19q13 and a segment of the homologous murine chromosome 7 which exhibit nine separate conserved linkage groups. The human and mouse X chromosomes contain a minimum of eight conserve syntenic groups.

Syntenic regions need not, however, necessarily be comparable in size. Thus, the human T-cell receptor β (TCRB; 7q35) region (800 kb in length) is considerably larger than its murine counterpart (500 kb) as a result of repeated duplications of specific Vβ segments in the primate lineage. Similarly, the human HLA class II region (6p21.3) at ~900 kb is approximately three times the length of its mouse equivalent as a result of the duplication of specific HLA-DP,-DQ and -DR members of this multigene family in the primate lineage. By contrast, the class III HLA regions of human and mice are remarkably similar in structure.

Many syntenic blocks extend across human centromeres. Thus, gene order in the pericentric regions of human chromosomes 1, 2, 4, 6, 11 and 19 is conserved in the homologous regions of murine chromosomes 3, 6, 5, 1, 2, and 8 respectively. The majority of rearrangements appears to be due to inversions within chromosomes rather than rearrangements between chromosomes but more detailed mapping data will be required to obtain the map resolution necessary for a definitive assessment. One *caveat* which should be borne in mind is that the linkage map may have been broken up to a large extent in rodents than in primates as compared to the ancestral mammalian genome. One way to study this type of rearrangement is by the comparative analysis of the chromosomal locations of the individual gene loci involved. For example, Maresco et al. (1998) determined the locations of the high-affinity immunoglobulin receptor genes (FCGR1A, 1q21; FCGR1B, 1p12; FCGR1C, 1q21) in the rhesus monkeys (*Macaca mulatta*), baboon (*Papio papio*) and chimpanzee (*Pan troglodytes*) thereby providing evidence for the occurrence of two pericentric inversions during the evolution of human chromosome 1.

Conservation of synteny may extent beyond the mammals. For example, 11/18 genes from the chicken Z chromosome have orthologues on human chromosome 9pter-q22, albeit in a different order. What are the reasons for this degree of conservation? One reason may have been functional, for example in order to allow coordinate regulation of the genes involved or in order to avoid possible meiotic disturbance consequent to a major chromosomal rearrangement. In the case of the HLA system, synteny may have indirectly promoted the generation of diversity by optimizing the potential for gene conversion. Alternatively, it is possible that insufficient time has passed for ancestral linkages to have been broken up completely. Synteny may be conserved in one phylogenetic group but not in another. Thus, the surfeit genes are tightly clustered in human (SURF1, SURF2, SURF4, SURF5; 9q34.1) and chicken but not in invertebrates where the Surf genes are unlinked.

Assuming a human-rodent divergence time of 80 Myrs, Collins and Jukes (1994) estimated the rate of silent substitution to be 2.9×10^{-9} site^{-1} year^{-1}. However, this is very much an average figure arrived at by comparison of the 4-fold degenerate sites in the protein-coding sequences of 337 human/rodent gene comparisons. Sequence conservation does vary quite considerably between proteins: indeed a study of 1196 orthologous mouse and human protein sequences revealed sequence conservation of between 36% and 100% with an average of 85%.

Large scale genomic DNA sequence comparisons between human and mouse are still difficult owing to the paucity of orthologous pairs of sequences > 20 kb in length. However, Koop (1995) noted three distinct patterns of sequence divergence in the noncoding DNA sequence of human and rodent genomes:

(i) A high level of sequence similarity in gene regions contrasting with divergent non-coding regions, for example β-globin (HBB; 11p 15.5) and γ-crystallin (CRYGA; 2q33-q35) genes.

(ii) A conserved pattern of sequence similarity noncoding regions, for example T-cell receptor-α (TCRA; 14q11.2) and -δ (TCRD; 14q11.2) genes and α- and β-myosin heavy chain (MYH6, MYH7; 14q11.2-q13) genes. At least some of this conservation may be attributed to regions which bind T-cell nuclear proteins which may play a role in the control of gene transcription.

(iii) A mixed pattern of sequence similarity, for example the immunoglobulin heavy chain J-Cμ-Cδ gene region (14q32.33): the J-Cμ portion exhibits ~64% sequence homology between human and mouse while the Cδ region shows little if any sequence conservation.

This 'mosaic model' of genome evolution may reflect differing rates of mutation or differential repair efficiencies between different regions of the genome. As the sequences of further syntenic regions become available [e.g. the Bruton's tyrosine kinase (BTK; Xq21.33-q22) gene region], this question can be addressed.

Primate Evolution

Adaptation and Adaptive Radiation

The order of primates originated in the Palaeocene some 50-60 Myrs ago and contains the monkeys, apes and humans. Originally adapted for arborial life, extant primates include low canopy runners (guenons), high canopy acrobats (spider monkeys) and the brachiating great apes whilst some have become exclusively terrestrial (baboons, mandrills, and humans). These habits are responsible for the adaptation of the skeleto-muscular system to allow jumping, swimming and grasping. Primates exhibit a wide variety of characteristics which have equipped them to exploit their various niches optimally. The Anthropoidea, with their relatively large brain size, possess binocular vision and high visual acuity but a relatively poor olfactory sense. Their high intelligence allows rapid and measured reaction to external stimuli. The extension of the time periods for gestation, developmental

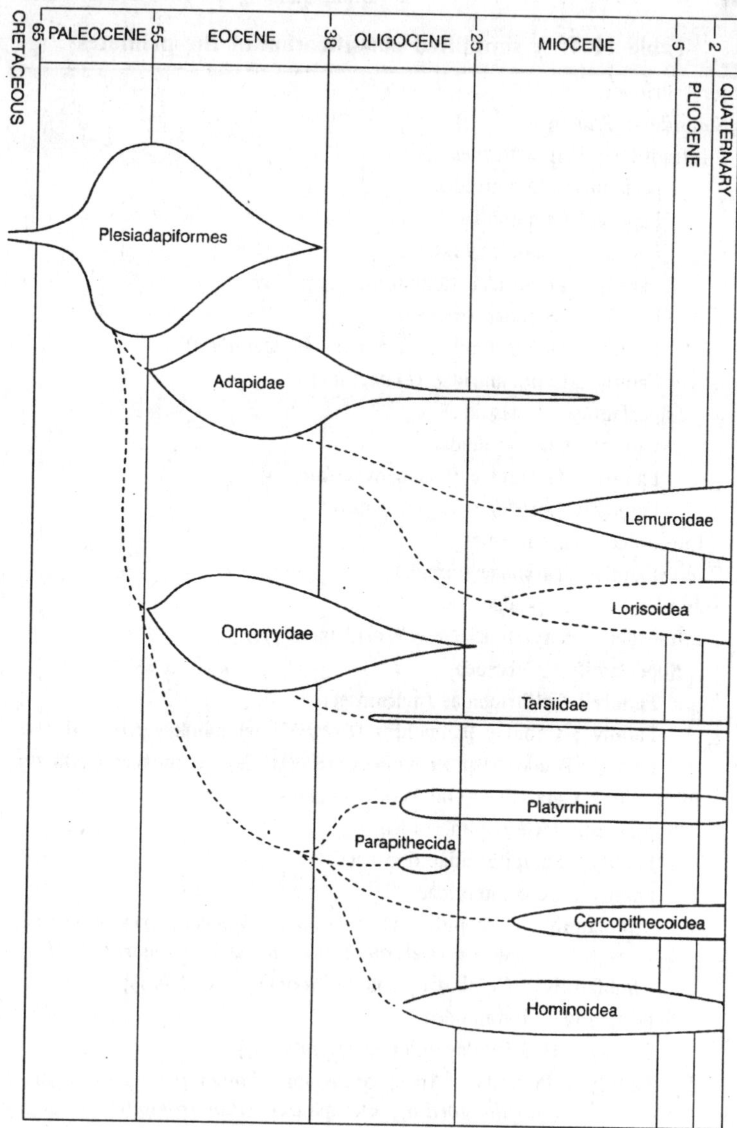

Fig. 11.2. Longevity of primate groups.

growth and parental care are associated with increased powers of learning whilst communication by use of elaborate vocal signals has allowed the development of complex patterns of social behaviour. The menstrual cycle with ovulation has been accompanied by the development

Table 11.4. A simplified classification of the primates

Order : Primates
  Suborder : Prosimii
    Infraorder : Lemuriformes
      Superfamily : Lemuroidea
        Family : Omomyidae
        Family : Adapidae (extinct)
        Family : Lemuridae (leumurs)
        Family : Indriidae (indris)
        Family : Daubentoniidae [aye-aye (*Daubentonia*)]
        Family : Lepilemuridae (*Lepilemur*)
      Superfamily : Lorisoidae
        Family : Cheirogaleidae
        Family : Galagidae [bushbaby (*Galago*)]
        Family : Lorisidae (*Loris*, potto)
    Infraorder : Tarsiiformes
        Family : Tarsiidae (tarsier)
  Suborder : Anthropoidae
    Infraorder : Platyrrhini (New World monkeys)
      Superfamily : Ceboidea
        Family : Callitrichidae (marmosets)
        Family : Cebidae [capuchins (*Cebus*), owl monkey (*Aotus*)]
        Family : Atelidae [spider monkeys (*Ateles*), howler monkey (*Alouatta*)]
    Infraorder : Catarrhini (Old World monkeys)
      Superfamily : Cercopithecoidea
        Family : Parapithecidae (extinct)
        Family : Cercopithecidae
          Subfamily : Cercopithecinae [macaques (*Macaca*), baboon (*Papio*), guenon (*Cercopithecus*), mandrill, langurs (*Presbytis*)
          Subfamily : Colobinae [colobus monkeys (*Colobus*)]
      Superfamily : Hominoidea
        Family : Hylobatidae [gibbon (*Hylobates*)]
        Family : Pongidae [Apes; orangutan (*Pongo pygmaeus*), gorilla (*Gorilla gorilla*), chimpanzee (*Pan troglodytes*), pigmy chimpanzee (*Pan paniscus*)]
        Family : Hominidae [Human (*Homo*)]

of sexual behaviour and signaling in the female. Finally, the omnivorous diet of primates may have been responsible for the development of their characteristics manual dexterity which has allowed them both to explore and manipulate their environment.

The order Primates contains about 200 living species which are divided into two suborders: the Prosimii and the Anthropoidea. The Prosimii, which originated in the Palaeocene have retained the insectivoran characteristics of long face, lateral eyes and small brain. The Anthropoidea comprise the Old World monkeys, the New World monkeys and the great apes. The New World or platyrrhine (flat nosed) monkeys are thought to have been isolated since the Eocene. The Old World or catarrhine monkeys share a common ancestor in the late Eocene and do not differ markedly in either habitats or organizations from the New World monkeys. The great apes include the gibbon and orangutan from East Asia and the chimpanzee and gorilla from Africa.

Primate Phylogeny

The phylogeny of the hominoid primates was initially investigated by means of DNA-DNA hybridization. These studies employed single copy nuclear DNA to calculate the temperature ($T_{50}H$) in degrees Celsius at which 50% of all single copy DNA sequences were in the hybrid form and 50% had dissociated. The delta $T_{50}H$ between chimpanzee and human is 1.6. Assuming a relationship of delta $T_{50}H$ = 1% base mismatches, this translates into ~3.2 × 10^7 mismatches between the chimpanzee and human genomes. Sibley and Ahlquist (1987)

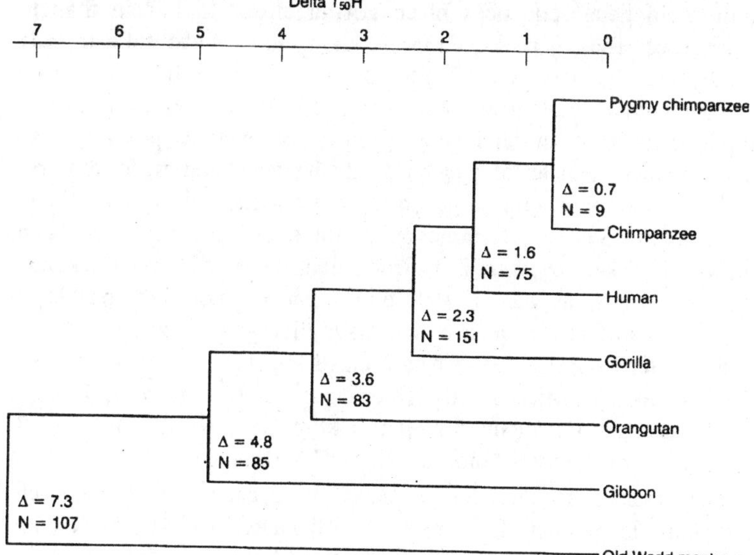

Fig. 11.3. Phylogeny of the hominoid primates as determined by average linkage clustering of delta $T_{50}H$ values derived from DNA-DNA hybridization.

estimated times of divergence for higher primates as: Old World monkeys, 25-34 Myrs ago; gibbons, 16.4-23 Myrs ago; orangutan, 12.2-17 Myrs ago; gorillas, 7.7-11 Myrs ago, chimpanzees-humans, 5.5-7.7 Myrs ago. It is now recognized that the chimpanzee is actually represented by two distinct species, the common chimpanzee (*Pan troglodytes*) and the pigmy chimpanzee (*Pan paniscus*) which diverged from each other ~2.3 Myrs ago.

The studies of Sibley and Ahlquist (1984, 1987) received support from Caccone and Powell (1989). However, the validity of conclusions drawn from DNA-DNA hybridization data has been challenged and the interpretation of these studies is still somewhat contentious. Sibley and Ahlquist's scheme is nevertheless in broad agreement with the fossil record, comparative morphology, immunological studies and chromosome phylogeny as well as being compatible with studies of individual DNA sequence. Thus, a similar picture has emerged from maximum parsimony analysis of DNA sequences derived from primate β-globin gene regions, the c-*myc* oncogene, the ribosomal DNA genes, the α-1,3-galactosyltransferase gene, the cytochrome P450 CYP21 gene and mitochondrial DNA as well as from transversion rates in nuclear and mitochondrial DNA.

However, divergence data from primate protamine P1 and α-fetoprotein gene sequences place gorilla closer to human than to chimpanzee. These gene sequences are known to have evolved extremely rapidly and the observed differences may well have been stochastic changes with no biological consequences for the species concerned or implications for their phylogeny. Similar explanations probably also apply to other examples of apparent gorilla-human closeness, for example studies of the mitochondrial genome and a polymorphism at the HLA-DQA locus. Studies of immunoglobulin ε pseudogenes have been equivocal. Taken together, it is clear is that the split between humans, chimps and gorillas was very close and that it is not unreasonable to expect that different sequences will have diverged to variable extents in the different lines. These data yielded slightly revised divergence times for higher primates with tighter error margins: gibbons, 14.6 ± 2.8 Myrs ago; orangutan, 8.2 ± 0.8 Myrs ago; gorilla, 6.7 ± 1.3 Myrs ago, chimpanzees-humans, 5.5 ± 0.2 Myrs ago.

The rates of accumulation of mutations appear to vary by as much as seven-fold between different primate lineages. The line of descent from the primate node to humans shows a slowdown in evolutionary rates from 7.7×10^{-9} fixed changes $site^{-1}$ $year^{-1}$ for the first 15 Myrs

(55-40 Myrs ago) to 1.3 $\times$ 10^{-9} for the next 15 Myrs (40-25 Myrs ago) 1.0 $\times$ 10^{-9} for the last 25 Myrs. The average evolutionary rate for the hominoids (1.1 $\times$ 10^{-9}) is lower than the rates for macaque, a catarrhine (1.9 $\times$ 10^{-9}) and for spider monkey, a platyrrhine (1.8 $\times$ 10^{-9}). By comparison, the line of descent from primate node to *Tarsius* shows an evolutionary rate of 3.4 $\times$ 10^{-9} fixed changes site^{-1} year^{-1} which is approximately half the stem-simian rate. The *hominoid slowdown* is at its greatest in human although anatomically, humans are quite divergent. Clearly, changes in certain key genes must have assumed a critical importance. As if perhaps to emphasize this point, some gene sequences appear to buck the trend; thus, the evolutionary rate of the noncoding region of the immunoglobulin-α gene is greater in hominoids than in Old World monkeys.

Chromosome Evolution in Primates

Old World primates

The evolution and probable phylogeny of primate chromosomes has been extensively reviewed by both Rumpler and Dutrillaux (1990) and Clemente et al. (1990). The interested reader is referred to these reviews for detailed accounts. Some chromosomes appear to have been relatively protected from change during primate evolution, for example human chromosomes 19 and X. By contrast, other chromosomes have been prone to significant reorganization, for example human chromosomes 1, 3, and 7.

The most frequent types of chromosomal change detected in primate evolution are inversions, changes in the amount and localization of heterochromatin, fusion and fissions, and changes in the location of centromeres due to activation/inactivation. Reciprocal translocations, deletions and insertions are much less frequent. Human chromosome 18 differs from the homologous chromosomes in the great apes by a pericentric inversion and it is thought that one inversion breakpoint may have been located at or within the centromere. Pericentric inversion breakpoints have also been identified on the chimpanzee equivalents of human chromosomes 4 (4p14, 4q21), 9 (9q22) and 12 (12p12 and 12q15) and these appear to coincide with the locations of either fragile sites or tumor-associated break-points. Pericentric inversions may have played an important role in establishing reproductive isolation and speciation during the evolution of the higher primates.

There are at least a dozen blocks of X-Y sequence homology outwith the pseudoautosomal region of humans but these blocks occur in a very different order and orientation on these chromosomes. This

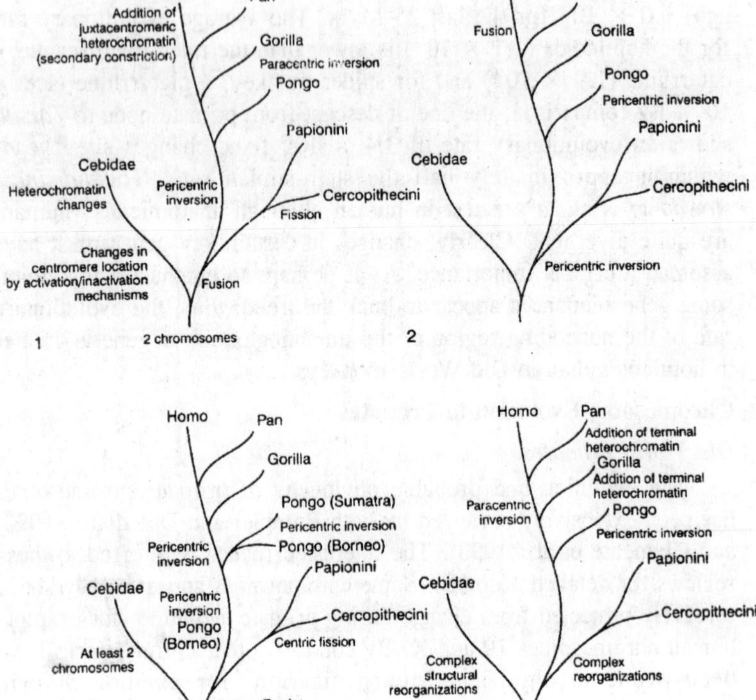

Fig. 11.4. Chromosome evolution in primates.

may be accounted for in terms of the occurrence of a number of different inversions, transpositions and other rearrangements during primate evolution. One example of a human-specific inversion is that involving the short arm of the Y chromosome. Sine humans from different racial groups all possess this Yp inversion, the rearrangement must have occurred prior to the divergence of the human racial groups.

Studies of chromosome banding patterns and hybridization homologies between ape and human chromosomes have provided evidence for human chromosome 2 having arisen form the fusion of two ancestral simian chromosomes showed that this probably occurred by telomere-telomere fusion at 2q13 rather than by translocation after chromosome breakage. This fusion has subsequently been confirmed by chromosome painting and since it accounts for the reduction in chromosome number from 24 pairs in the great apes (chimpanzee, orangutan, and gorilla) to 23 pairs in humans, it must have been a relatively recent event. Fusion must have been accompanied or followed by inactivation or removal of one of the ancestral centromeres. Consistent with this postulate, IJdo et al. (1991) found evidence by hybridization for the residual presence of an ancestral centromere at 2q21. Clearly, this reduction in chromosome number would have been a critical event during the speciation process; if it was not in itself responsible for bringing about reproductive isolation, it would certainly have helped to maintain it.

Another major chromosomal rearrangement to have occurred in the great apes is to be found in the gorilla. Using human chromosome-specific libraries as probes for *in situ* hybridization, Stanyon et al (1992) described a reciprocal translocation in the gorilla lineage which is not present in the chimpanzee. Thus, chromosomes 4 and 19 of the gorillas were derived from a reciprocal translocation between the ancestral chromosomes homologous to human chromosomes 5 and 17. Wienberg et al. (1990) used chromosomal *in situ* suppression hybridization to demonstrate that the centromere of human chromosome 17, the long arm of the chromosome and a small part of the short arm

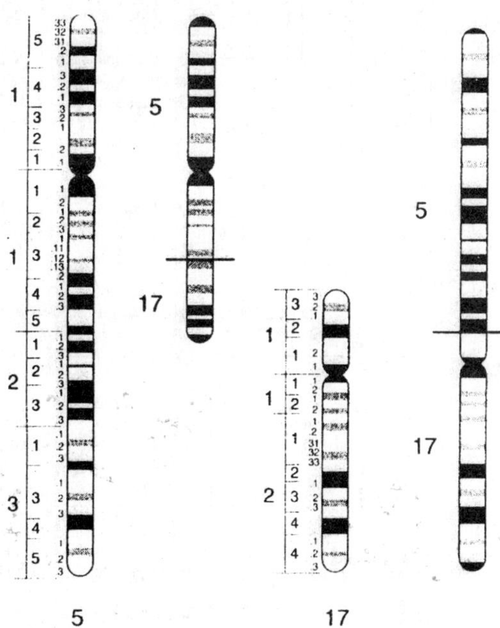

Fig. 11.5. Idiograms of human chromosome 5, gorilla chromosome 19, human chromosome 17, and gorilla chromosome 4.

all contribute to gorilla chromosome 4 whilst most of the short arm of chromosome 17 contributes to gorilla chromosome 19.

Probably the best understood human chromosome in terms of its evolution is chromosome 21. The equivalent of human chromosome 21 (HSA21) formed a large and unique chromosome together with chromosome 3 (HSA3) in the eutherian ancestor. This chromosome was conserved without significant alterations only in lemurs, the civet and the pig. It underwent inversions in the tree shrew and the cow. Various translocations involving the portion corresponding to HSA3 occurred in the brown lemur, cat, rabbit and mouse. In the primates, two independent fissions occurred. In New World monkeys, a small segment of HSA3 remained attached to SHA21 and this chromosome

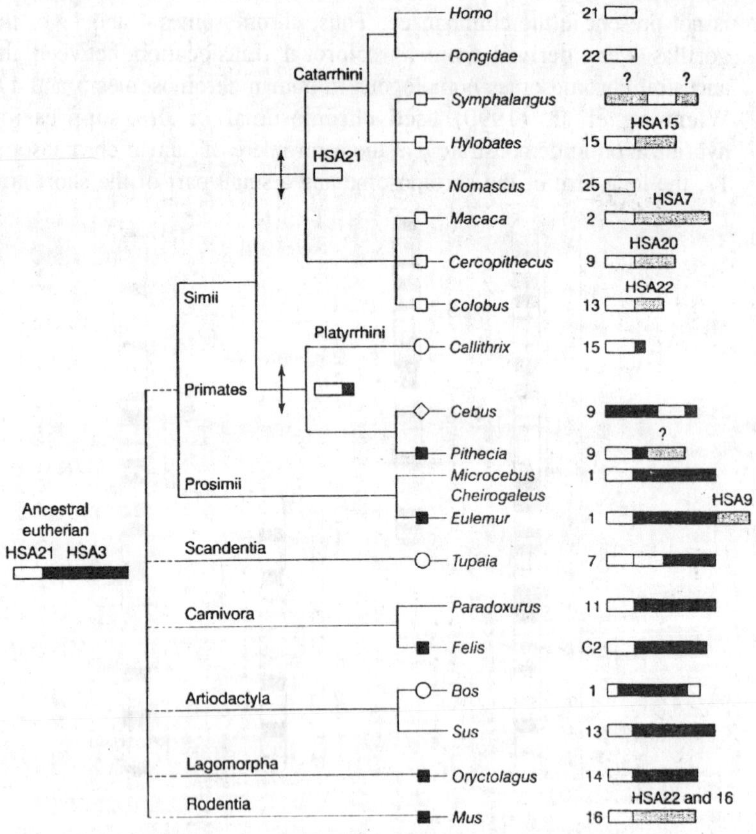

Fig. 11.6. Evolution of the equivalent of human chromosome 21 (HSA21) in eutherian mammals.

then underwent further rearrangements; an inversion in the marmoset, the addition of heterochromatin in the capuchin monkey and a translocation in the saki monkey. HSA21 was formed in the common ancestor of Old World monkeys and underwent translocations with various equivalents of human chromosomes in all the Cercopithecidae. HSA21 was conserved without visible alteration in the black gibbon and the great apes.

One technique which is proving extremely useful in primate cytological studies in cross-species chromosome painting (CSCP; also known as comparative painting or ZOO-FISH). CSCP involves the hybridization of a chromosome-specific paint from one species (usually human) onto metaphase spreads of another species. This approach allows the rapid construction of chromosome maps from the primate species in question which can reveal cytogenetic homologies with the human karyotype. By these means, a high degree of synteny has been found between human and baboon chromosomes. By contrast, numerous translocations are apparent in the gibbon (*Hylobates lar*, 2n = 44) genome as compared to human and the great apes; the 22 human autosomes have to be divided into 51 elements in order to recombine them into the 21 gibbon autosomes. Similarly, in the Concolor gibbon (*Hylobates concolor*, 2n = 52), the 22 human autosomes have to be divided into 63-67 segments in order to recombine them into the 25 gibbon autosomes.

Despite the degree of genetic similarity between the great apes, the chimpanzee genome is approximately 10% larger than that of the human of gorilla. Nevertheless, the structure of the human and chimpanzee genomes is highly conserved at both chromosomal and sub-chromosomal levels. Even at lower resolution, the extent of evolutionary conservation is readily apparent. Human microsatellite DNA sequences are sufficiently well conserved in chimpanzees that human PCR primers can be used to amplify $(CA)_n$ repeats in the chimpanzee. Differences in microsatellite allele length between humans and other primates have however been noted. Crouau-Roy et al. (1996) studied microsatellites within a 30 cM region of human chromosome 4p and found that all informative loci which are linked in human were also linked in the chimpanzee, indicating that evolutionary conservation extends to the locus level. In general, heterozygosity was found to be greater in chimpanzees, a reflection perhaps of the greater genetic diversity in chimpanzee populations. Some loci, however, appeared to be less heterozygous than in human, a phenomenon that appears to be caused by interruptions of the repeat elements at these loci.

Human chromosomes are not always identical. Indeed many exhibit heteromorphism, especially in the centromeric and satellite regions of the acrocentric chromosomes. Chromosomes 1, 9, 13, 14, 15, 16, 19, 21, 22, and Y are the most heteromorphic whilst chromosomes 2-8 and X are the least heteromorphic. Inter-chromosomal variation can be substantial; two homologues of chromosome 21 having been noted to vary in healthy individuals by as much as 21 Mb or 40% of the length of the chromosome. Family studies have shown that such heteromorphisms are not artefactual and can be inherited in mendelian fashion. It would appear that this variation can be largely ascribed to variation in the size of repeat sequence arrays and probably results from unequal crossing over between different classes of repetitive element. Bivariate flow karyotyping has been used to study the relative DNA content of homologous chromosome pairs in individuals from different racial groups. Significant variation in DNA content, ranging from 10 to 40%, was found for chromosomes 1, 13, 14, 15, 16, 19, 21, 22, and Y. However, the spectrum of variation observed in the different racial groups was very similar.

New World primates

Like gibbons, many New World primates possess highly rearranged genomes. All new World primates have a translocation between chromosomes 8 and 18. The Cebidae possess translocations between chromosomes 10/16 and 2/16 whilst the Atelidae are characterized by translocations between 3/15 and 4/15. Significant synteny is nevertheless apparent between human and the capuchin monkeys, *Cebus*, between human and the Colobus monkey, between human and the howler monkey, between human and the black-handed spider monkey, *Ateles geoffroyi* and between human and the marmoset *Callithrix*.

Evolution of the Human Sex Chromosomes and the Pseudoautosomal Regions

The human sex chromosomes are heteromorphic; the X chromosome contains ~160 Mb DNA and perhaps 3000 genes whereas the Y chromosome contains only 60 Mb DNA and probably only a handful of genes. The Y chromosome is largely composed of constitutive heterochromatin harboring different families of repetitive DNA. Despite the size difference between the sex chromosomes, they are nevertheless able to pair successfully during meiosis. Recombination, however, is largely confined to the two *pseudoautosomal regions* (PARs); a major PAR (2.6 Mb in size) at the tips of the short arms of X and Y chromosomes which is the site of an obligate crossover during male meiosis, and a minor PAR (320 kb) at the tips of the long arms of the X and Y chromosomes. Regular X-Y recombinational exchanges have served to maintain homology between the chromosomes in the PAR regions.

Gene mapping studies have shown that part of the eutherian (placental) mammalian X chromosome ('conserved region'; XCR) is shared by the X chromosomes of marsupials and monotremes. Since a series of genes on the short arm of the human X are clustered in two autosomal groups in marsupials and monotremes, these loci define a region (XRA) that has been recently added to the X chromosome in eutherian mammals. Thus, the X chromosome of the common mammalian ancestor was smaller than that found in extant eutherians and at least two autosomal regions have since been added to it. The location of this X-autosome fusion corresponds to the border between the XCR and the XRA and lies at Xp11.23.

The human X and Y chromosomes also exhibit substantial homology outwith the pseudoautosomal region. Although this homology is consistent with these chromosomes having once constituted a homomorphic pair, the observed homologies also reflect intra-chromosomal duplication events followed by inter-chromosomal translocations. For example, the

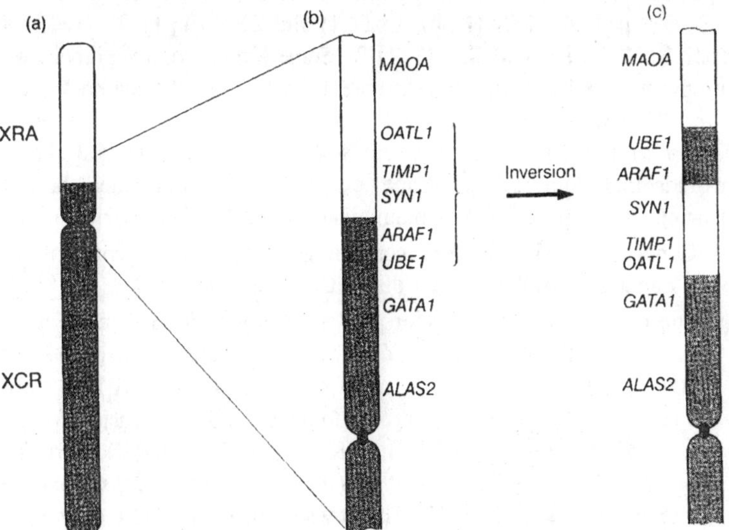

Fig. 11.7. (a) Evolutionary origins of the human X chromosome. XCR, conserved region of the X. XRA, recently added region. (b) Details of original gene order at the fusion point. (c) Possible inversion event that placed the XCR genes, UBE1 and ARAF1, between the XRA genes, SYN1 and MAOA.

minor PAR is thought to have originated during human evolution by the translocation of 320 kb of X chromosomal sequence to the Y chromosome, an event which may have been mediated by recombination between LINE elements.

The sex chromosomes however vary dramatically in terms of their evolutionary conservation. The Y chromosome has undergone significant changes during mammalian evolution. By contrast, the X chromosome exhibits conservation of synteny between human and mouse. This is considered to be a sequence of X inactivation because X-autosome translocations would have tended to be disadvantageous by virtue of their interference with the dosage compensation mechanism. Despite this, gene order on the X chromosome has changed significantly between mouse and human with numerous inversions altering the relative position of genes.

The human Y chromosome contains a number of active genes. Most notably it hosts the rapidly evolving sex determining gene, SRY (Yp11.3) which occurs in the sex chromosome-specific region, about 5 kb from its boundary with the major PAR. In human, a few X-linked genes have functional homologues on the Y chromosome (e.g. CSF2RA (Xp22.32, Yp11.3), MIC2 (Xp22.23, Yp11.3), RPS4X (Xq31.1) and RPS4Y (Yp11.3), ZFX (Xp21.3-p22.1) and ZFY (Yp11.3), AMELX (Xp22.1-p22.31) and AMELY (Yp11.2). Some Y chromosome homologues are however nonfunctional, for example the KAL1 pseudogene on Yq11.2 or the XG pseudogene on Yq11.21. One consequence of the sequence identity of the PARs between the sex chromosomes is that the X chromosomal homologues escape X inactivation in females thereby ensuring that gene dosage is maintained at the level found in males.

Genes identified in the nonrecombining portion of the Y chromosome (NRY) appear to fall into two distinct classes.

(i) Those that are specifically or predominantly expressed in the testis [sex determining gene (SRY; Yp11.3), deleted in azoospermia (DAZ; Yq11), RNA-binding motif protein 1 (RBM1; Yq11), testis-specific protein (TSPY), chromodomain Y (CDY), basic proteins Y1 and Y2 (BPY1, BPY2), XK related Y (XKRY), tyrosine phosphatase PTP-BL related Y (PRY) and testis transcripts Y1 and Y2 (TTY1, TTY2)]. That these genes have not only been retained on the Y chromosome, but in two cases have also been amplified, may have been part of an evolutionary strategy to optimize male reproductive fitness.

(ii) Those that are widely or ubiquitously expressed and which have closely related counterparts on the X chromosome [dead box Y (DBY), thymosin β4 (TB4Y), translation initiation factor 1A (EIF1AY), ubiquitous TPR motif Y (UTY) and *Drosophila* fat facets-related (DFFRY; Yp11.2), AMELY (Yp11.2), RPS4Y (Yp11.3), zinc finger protein Y (ZFY; Yp11.3) and SMCY]. Conservation of specific X–Y gene pairs may have been associated with a requirement to maintain comparable expression levels for certain housekeeping genes between males and females. Consistent with the predictions of this postulate, the X chromosome homologues of these Y-borne genes escape X inactivation.

The mammalian sex chromosomes are thought to be descended from a homologous pair of autosomes. This process could have been initiated with the evolutionary appearance of the testis-determining SRY gene on the nascent Y chromosome, probably by duplication, translocation and subsequent divergence of the X-linked SOX3 (Xq26-q27) gene. Suppression of recombination with its homologous chromosome (the nascent X) then led to the gradual degeneration of the Y chromosome owing to its inability to segregate genes carrying deleterious alleles. Evidence for this degenerative process comes from several sources. The rate of nucleotide substitution in Y chromosome genes appears to be ~2-fold higher than the rate exhibited by X chromosomal genes although the frequency of DNA sequence polymorphism may be lower in the sex-specific region of the Y chromosome than in the PARs. The Y chromosome also exhibits a high frequency of retroviral insertion in humans, chimpanzees and orangutans. Finally, there is emerging evidence for gene loss from the Y chromosome during mammalian evolution. In mouse and human, the ubiquitin-activating enzyme (UBE1) gene is located on the X chromosome (Xp11.23-p11.3). A copy of the *Ube1* gene is also located on the Y chromosome in the mouse, ring-tailed lemur, squirrel monkey (*Saimiri sciureus*) and marmoset (*Callithrix jacchus*) but not in the Old World monkeys, chimpanzee or human, indicating loss of the Y-linked gene >35 Myrs ago during the evolution of the primates. Similarly, the Y-linked copy of the EIF2S3 gene (Xp22.1-p22.2) encoding the eukaryotic translation initiation factor EIF-2γ was lost 35-60 Myrs ago in a common ancestor of the simian primates. The gradual degeneration of the Y-chromosome implies that the retention of functional gene copies on this chromosome for a significant period of evolutionary time could have conferred some selective advantage.

It should however be noted that the Y chromosome can also acquire genetic material from other chromosomes. For example, the multicopy DAZL1 gene (Yq11.23; deleted in azoospermia) was transposed to the Y chromosome from an autosome during primate evolution as was the multicopy RNA-binding motif (RBM1; Yq11) gene. It may be that transfer to a male-specific location provided protection against inactivation or loss. Another example of the duplicational transposition of a gene to the Y chromosome is that of AMELX (Xp22.1-p22.31) and its Y-chromosome counterpart AMELY (Yp11.2); the latter gene, which appears to be fully functional, is present on the Y-chromosomes of bovids and primates but not rodents thereby dating the transpositional event to at least 40 Myrs ago.

In humans, the XG blood group gene (Xp22.32) spans the major PAR on the X chromosome—the first three exons are pseudoautosomal whereas the remaining seven are X chromosome-specific. In humans and the great apes, an *Alu* sequence is located at the boundary between the major PAR and the Y chromosome-specific DNA but this sequence is not present in Old World monkeys. The *Alu* sequence was therefore inserted into the pre-existing boundary after the divergence of the great apes from the Old World monkeys. Although it did not create the boundary, the *Alu* sequence does serve to demarcate it.

Ellis et al. (1994) proposed a model for the formation of the boundary of the major PAR. They hypothesized a pericentric inversion of the Y chromosome with one breakpoint in the ancestral XG gene and the other breakpoint 5 kb distal to the ancestral SRY gene. In a refinement of this postulate, Fukagawa et al. (1996) suggested that the inversion occurred by illegitimate recombination between two PAR boundary sequences, one in the ancestral XG gene and the other near the ancestral SRY gene.

The PAR has undergone quite rapid change during mammalian evolution involving both gene duplication and translocation events in the region (e.g. STS, MIC2, XG, CSF2RA, IL3RA, ARSD, ARSE) and resulting in the movement of the PAR boundary to create X-unique regions. The evolution of the PAR and the divergence of the mammalian X and Y chromosomes may be viewed in terms of the 'addition-attrition' hypothesis. This states that the incorporation of autosomal sequences into the PAR of either the X or Y chromosome initially served to generate homologous regions which could pair at meiosis. Recombination with an homologous partner could then result in PAR enlargement. Alternatively, the steadily accumulating mutations on the Y chromosome would have served to decrease the level of homology to the X chromosome thereby reducing PAR size. Fukagawa et al. (1996) proposed a further twist to this argument in that once divergence had reached a certain level, recombination frequency would have decreased thereby further increasing the rate of divergence.

Evidence in favour of the addition-attrition theory comes from the dynamic nature of the major PAR region during mammalian evolution. Thus, the STS gene which is X-linked in humans and the great apes (Xp22.32) is autosomal in prosimians as the ANT3 gene which is pseudoautosomal (Yp11.3) in humans. Similarly, the human pseudoautosomal genes CSF2RA (Xp22.32/Yp11.3), SHOX (Xp22/Yp11.3) and IL3RA (Xp22.3/Yp11.3), are autosomal in the mouse. Further evidence for the process of attrition may come from the finding that whereas the *Fxy* gene spans the PAR on the murine X chromosome, its human counterpart (FXY; Xp22.3) lies proximal to the human PAR.

The 'X-driven' hypothesis of Graves (1995; 1998) essentially proposes that the rapid evolutionary spreading of X inactivation preceded the decay of Y chromosomal genes and even drove its initial steps. This hypothesis predicts that inactivated X-linked genes with functionally comparable Y-linked homologues should exist as evolutionary intermediates, but as yet no such gene has been found in any mammalian species. Indeed, a considerable number of human X-linked genes escape X-inactivation but have no detectable Y-borne counterpart.

An alternative 'Y-driven' pathway of X-Y gene evolution has been proposed by Jegalian and Page (1998). Briefly, these authors suggested that many extant genes represent intermediates on a general pathway by which X-Y genes or gene clusters evolved form autosomal genes. Autosomal genes would have entered the pathway either by virtue of their presence on the emergent sex chromosomes or via translocation of an autosomal gene. This would have been followed by suppression of X-Y recombination. These steps occurred either at the chromosomal or sub-chromosomal level and gave rise to functionally equivalent X-linked (but not inactivated) genes and Y-linked genes. Subsequently, three different processes (Y gene decay, upregulation of X-linked gene expression, and X-inactivation) interacted resulting in an inactivated X-linked gene accompanied by the loss of the Y gene. Expression of the X-linked gene then increased as an adaptation to the reduced or restricted expression of its Y-linked counterpart. This compensated for the loss of Y gene function and restored optimal expression levels in males. X-inactivation, on the other hand, may be viewed as a counter-

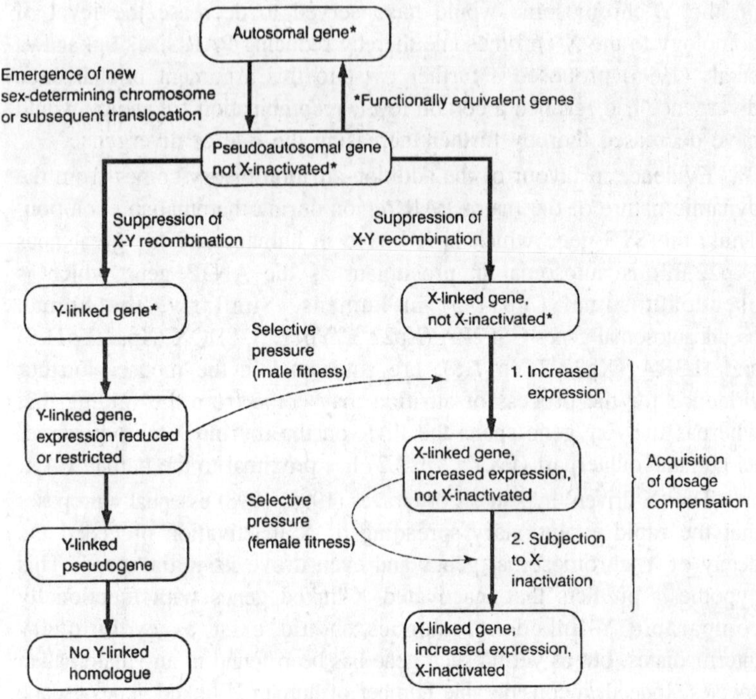

Fig. 11.8. A proposed Y-driven pathway for X-Y gene evolution in mammals.

response which restored optimal expression levels in females. This explanation could in principle account for most X-linked and X-Y homologous genes in extant mammals, many of which exist at intermediate steps in the pathway. Jeglian and Page (1998) pointed out that only one gene cannot be accommodated in their pathway schema: the human pseudoautosomal gene, SUBL1, which is X-inactivated and transcriptionally silenced on the Y chromosome.

Evolution of the Mitochondrial Genome

The 16 569 bp of the human mitochondrial genome encodes 13 polypeptides, all subunits of the enzyme complexes of the pathway of oxidative phosphorylation, and a total of 22 tRNAs. The mitochondrial genome is characterized by its high proportion of coding DNA, the paucity of repetitive DNA sequence, the absence of introns within its genes and its own genetic code distinct from that of the nuclear genome. Mitochondrial genes experience a mutation rate that has been estimated to be up to 17 times higher than the corresponding rate for nuclear genes. This is thought to be due to the fact that the mitochondrial genome undergoes many more rounds of replication but may also be a consequence of the relative error proneness of the mitochondrial DNA polymerase γ and increased exposure to the potentially mutagenic products of oxidative metabolism. Whatever the explanation, one consequence of the higher mutation rate has been that the mitochondrially encoded subunits of the oxidative phosphorylation enzyme complexes have evolved at a much higher rate than their nuclear encoded counterparts.

The origin of mitochondria is thought to have been through endocytosis by an anaerobic eukaryote of an aerobic eubacterium possessing on oxidative phosphorylation system. During the evolutionary transformation of endosymbiont to organelle, there has been significant transfer of DNA sequences from the mitochondrial to the nuclear genome. Interestingly, the mitochondrial genome of the slime mould *Dictyostelium* has retained a gene encoding a NADH: ubiquinone oxidoreductase subunit which was transferred to the nuclear genome in the common ancestor of other eukaryotes. Some mitochondrial DNA sequences present in the human genome as pseudogenes have been incorporated only relatively recently in primate evolution. It has been suggested that coevolution may have occurred between nuclear and mitochondrial genes, for example, cytochrome c oxidase subunit IV (COX4; 16q22-qter and mt genome).

Mitochondrial DNA variation has been very informative for the study of the evolution of human population. The interested reader is referred to Stoneking (1996) for a review of the topic.

Evolution of Human Populations

Discussion of the fossil record of early hominids is outwith the remit of this volume and the interested reader is referred to Wood (1992, 1996) for readable preces. Similarly, the origin of modern humans and the movement of human populations have been well reviewed by a number of authors including Cavalli-Sforza et al. (1994), Cavalli-Sforza (1998), von Haeseler et al. (1995), Jones, Martin, and Pilbeam (1992) and Lewin (1998).

There are currently two different views of the origin of modern humans. The first, which is not inconsistent with the fossil record, proposes that different populations ('races') of *Homo-sapiens* evolved independently from their ancestor *Homo erectus* in different parts of the Old World ('multiregional model'). Migration of *H. erectus* from Africa to the rest of the Old World may have occurred ~ 1 Myrs ago. The alternative hypothesis is that *H. sapiens* arose once in Africa and that the species may have been subjected at some stage to a severe population bottleneck. With both models, the geographical separation of populations has led to the emergence of morphological differences although continued gene flow between populations has served to ensure that human gene diversity remains graded rather than discrete.

One of the most dramatic demonstrations so far of the utility of molecular genetics to the study of human gene evolution has been the analysis of mitochondrial DNA (mtDNA) from a fossil Neanderthal specimen. Neanderthals and humans are considered to have shared a common ancestor between 550,000 and 690,000 years ago. Since the Neanderthal mtDNA sequence was found to be quite distinct from those of modern humans, it would appear as if Neanderthals became extinct without contributing mtDNA to human populations, a finding consistent with the Out of Africa hypothesis.

The question of the age of the human gene pool has been approached by studying variation associated either with the mitochondrial genome or with Y chromosome-derived DNA sequences (mtDNA exhibits a high rate of evolutionary change and, along with most of the Y chromosome, is transmitted without recombination). Studies of mtDNA have yielded dates of the order of 200,000 years ago for the origin of modern humans, broadly consistent with estimates derived from Y-chromosome-derived DNA sequence data. Use of intronic variation in the ZFX (Xp21.3-p22.2) gene yielded a figure of 306,000 years that is almost certainly discrepant since the 95% confidence interval was extremely broad. Using 30 microsatellite loci to construct a phylogenetic tree for 14 different human populations, Goldstein et al. (1995) estimated that the time since divergence of African and non-African populations was 156,000 years. The above estimates are broadly compatible with those derived from polymorphisms associated with the CD4 (12p12-pter) gene, microsatellite data and protein polymorphism data. It must be remembered, however, that these results simply reflect the time elapsed since the most recent common ancestor for the sample population rather than the most recent common ancestor of all humans. Small populations and/or population bottlenecks will have served to obscure the actual timing of the 'origin' of modern humans.

As to the place of origin of modern humans, Wainscoat et al. (1986) claimed that the relative frequencies of haplotypes of the β-globin (HBB; 11p15.5) gene is African and nonAfrican population provided evidence for a migration out of Africa by a fairly small population. Further evidence for a recent African origin for modern humans comes from the observation that African populations have a ~20% greater microsatellite sequence diversity as compared with Asian and European populations. Studies of mitochondrial DNA have also shown that there is greater genetic diversity between African populations than among Asian or European populations. Indeed, these authors showed that genetic variation among humans on all continents are subsets of the variation present in Africans. However, some polymorphism lineages do not show deep branches for African populations which has made the Out of Africa hypothesis somewhat contentious. The higher level of genetic diversity manifested by African populations may simply be a reflection of their greater population size over the last million years.

Y chromosome variants appear to be more highly clustered geographically than those of mtDNA. One explanation for this difference could be that male migration has been more limited than that of women.

About 84% of human genetic diversity exists as differences between individuals within populations but the remaining 16% can be used to distinguish between populations. By comparison with apes, the extent of the genetic variation exhibited by modern humans is relatively low. This lack of genetic diversity is likely to be reflection of long-term small population size [before the introduction of agriculture 10,000 years ago, the entire human population probably did not exceed 100,000 and is thought to have been around 10,000 for most of its history], the effects of past population bottlenecks and the explosive population growth particularly during the last 10,000 years.

In human history, demographic expansions have often occurred as a result of technological developments affecting food availability and transportation fuelled by the pursuit of military or economic objectives. In terms of establishing genetic differences between populations, linguistic barriers have also been important.

Action of Natural Selection in Human Populations

Evidence for the recent effects of natural selection on human populations comes indirectly from observed variation in allele frequencies with infectious disease often serving as the selecting agent. The classical example of pathogen-driven selection is the heterozygote advantage accruing to carries of the Glu6→Val sickle cell mutation in the β-globin (HBB) gene which confers resistance to *Falciparum* malaria. It has been suggested that the high frequency of certain disease in specific populations may be explained in similar ways.

Heterozygote advantage has been invoked to account for the spread of the common cystic fibrosis mutation (δF508) in the CFTR (7q31.3) gene in Caucasian populations. The basis for heterozygote advantage was proposed to be increased fitness of heterozygous carriers during cholera epidemics. However, it is difficult to see how such overdominant selection could have brought about the extremely high prevalence of one specific CFTR lesion (δF508) relative to the large number of alternative CFTR mutations known. It would also be difficult to explain the gradient in δF508 frequency across Europe unless there was also a gradient of selective pressure. Thus, the most parsimonious explanation is probably genetic drift.

A large body of data has accumulated on HLA polymorphism and its relationship to disease susceptibility, resistance and progression. Polymorphic alleles at both HLA class I and II loci have been shown to be under selection. One recent example is selection for specific HLA class II (HLA-DR and HLA-DQB; 6p21.3) alleles as a result of hepatitis B virus infection. More than 90% of the 135 known HLA-DRB1 alleles appear to have been generated since the divergence of human and chimpanzee such changes appear to have arisen both by point mutation and by gene conversion and are consistent with the existence of substantial selective pressure.

Pathogen-driven selection may also have been responsible for increasing the frequency of genetic variants at other loci, for example the CCR2 (3p21) and CCR5 (3p21) chemokine receptor genes and the stromal cell-derived (SDF1; 10q11.2) gene. With the advent of the HIV epidermic, these variants may become advantageous in that they appear able to restrict HIV-1 infection and decrease the progression of HIV-1 infections to AIDS. Genetic susceptibility to parasitic infections may be under oligogenic control (e.g. *Leishmania*, *Schistosoma mansoni*, *Mycobacterium tuberculosis* and *Plasmodium falciparum*) implying that a number of different variants at different loci may be subject to selection. Infectious and parasitic disease is currently estimated to kill up to 20 million people in the world every year with acute respiratory infection, tuberculosis, diarrhoea, malaria, measles, hepatitis B, whooping cough and tetanus responsible for ¾ of this toll. Clearly, it is likely that selection will continue to operate on existing genetic variation at a wide range of genetic loci that serve to determine both the host's susceptibility and resistance to the disease in question.

Polymorphism of the human ABO blood group system (ABO; 9p34) may owe its origins to balancing (overdominant) selection mediated by infectious agents. Intriguingly, the same substitutions that differentiate the A and B alleles in humans are also present in the great apes and Old World monkeys leading to speculation that these polymorphic antigens may have arisen early in primate evolution. However, intronic sequence data point instead to an origin for the human alleles about 4 Myrs ago which argue for a model of convergent evolution of the blood group antigens in the higher primates.

Whilst pathogen-driven selection may be an important factor in increasing the frequency of certain genetic variants at specific human loci, susceptibility to infectious disease is unlikely to be the sole means by which natural selection influences allele frequencies. One common genetic variant not thought to be associated with infectious disease has been found in factor VII. Plasma levels of factor VII vary significantly in the general population and are known to be influenced by a number of different environmental factors including sex, age, cholesterol, and triglyceride. An Arg/Gln polymorphism at residue 353 of factor VII (F7; 13q34) which occurs with a frequency of about 10% in various populations, is associated with a 20-25% reduction in the level of plasma factor VII activity as a result of the impaired secretion of Gln variant. This high frequency is suggestive of a balanced polymorphism and could indicate that the Gln variant confers some benefit, for example protection against thrombosis, myocardial infarction or arterial disease. In support of this postulate, Silveira et al. (1994) have shown that the Gln allele is associated with a reduction in the amount of activated factor VII (FVIIa) generated in response to fat intake; individuals with the Arg/Gln genotype were found to possess FVIIa levels 48% of that exhibited by individuals homozygous for the Arg allele. Interestingly, a decanucleotide insertion polymorphism at -323 in the F7 gene promoter has been shown to be associated with a 33% reduction in promoter activity *in vitro* and a lower level of plasma factor VII activity and antigen *in vivo*. The occurrence of two functionally significant polymorphisms in the same gene is consistent with the idea of a selective advantage accruing to individuals with reduced factor VII activity.

A similar hemostatic system polymorphism is factor V Leiden. This variant, which underlies the phenomenon of activated protein C resistance, results from the substitution of Arg506 by Gln in coagulation factor V (F5; 1q23). Factor Va serves as a cofactor in the activation of prothrombin by factor Xa and the factor V Leiden variant is relatively resistant to activated protein C-mediated inactivation. Between

1% and 7% of the Caucasian population possess the factor V Leiden mutation which may therefore be regarded as a fairly frequent polymorphism with phenotypic effect. Since the factor V Leiden mutation is also associated with a relative risk of ~6.0 for venous thrombosis, this is also a polymorphism with clinical effect. Why is this factor V variant so common? Its high frequency in the general population suggests that it confers, or has conferred, some selective advantage on its bearers. Dahlback (1994) speculated that a slight hypercoagulable state associated with possession of the factor V Leiden variant might have been advantageous in certain situations such as traumatic injury and childbirth. Consistent with this postulate, Lindqvist et al. (1998) have shown that carries of the factor V Leiden variant have a significantly reduced risk of bleeding during childbirth. It may be that other common polymorphic variants in hemostatic factor genes e.g. the G20210A transition in the 3' untranslated region of the prothrombin (F2; 11p11-q12) gene, are explicable by similar models.

Other examples of polymorphic variants which may have conferred a selective advantage on carriers are the 'insertion' (I) allele of the angiotensin-converting enzyme (DCP1; 17q23) gene which appears to be associated with improved human endurance and the Glu487/Lys polymorphism in the aldehyde dehydrogenase 2 (ALDH2; 12q24) gene which is associated with alcohol sensitivity and alcohol avoidance. Selection is likely to have also operated on a variety of other human characteristics including cognitive ability, both form, skin pigmentation and pharmacogenetic variation.

Sequencing the Genomes of Model Organisms and Humans

The characterization of the genomes of a number of different and disparate species should aid significantly our understanding of the human genome, its structure, function and evolution. Such species ('model organisms') include a bacterium (*Escherichia coli*), a yeast (*Saccharomyces cerevisiae*), a nematode (*Caenorhabditis elegans*), the fruitfly (*Drosophila melanogaster*), the pufferfish (*Fugu rubripes*), the mouse, and the rate. Sequence o the genomes of these model organism is essential for the discovery, description and characterization of all genes within those genomes and proteins that these genes encode. It will provide information not only on the chromosomal organization of genes and gene families but also on the elements.

The 4.64 Mb genome of *E. coli* has been sequenced and encodes 4288 protein coding genes. By comparison, the 12.1 Mb genome of *S. cerevisiae*, which is organized into 16 chromosomes, contains 5885 protein-coding genes, ~140 ribosomal RNA genes, 40 snRNA genes and 275 tRNA genes.

The entire 97 Mb genome of *C. elegans* has also been sequenced and represents the first fully characterized genome of a multicellular eukaryote (*C. elegans*). This sequence predicts a total of 19,099 protein-coding genes and at least several hundred further genes specifying noncoding RNAs. At least 36% of *C. elegans* proteins exhibit a match in humans whilst 74% of characterized human proteins exhibit a match with a *C. elegans* protein. Each *C. elegans* gene has an average of five introns and the exons constitute some 27% of the nematode genome.

Comparison of the complete gene/protein sets of yeast and nematode has revealed that for a substantial proportion of the two organisms' genes, one-to-one orthologous relationships are identifiable. This suggests that the functions of many gene products were already established in the common ancestor of fungi and the metazoa. By contrast, most of the *C. elegans* signaling and regulatory genes that are known or expected to be involved in multicellularity have no yeast orthologue even though they may contain domain sequences present that are in yeast.

The possession of the compete genome sequences of various model organisms is proving of enormous benefit in identifying the human homologues of genes that are shared between these organisms and humans. Thus, the expressed sequence tags (EST) database (dbEST) can be screened using model organism genes as 'probes'. An example of this approach (termed 'cyberscreening' or '*in silico* cloning') is provided by the cloning of five human orthologues of yeast genes encoding proteins of the mitochondrial respiratory chain complex.

Within the next 5 years, the 3200 Mb human genome sequence should also become available. This will permit integration of cytogenetic, genetic, physical and transcriptional maps of the genome, information on inter-individual polymorphic variation, and the genotype-phenotype relationship particularly in the context of complex traits. The availability of the human genome sequence will lead to the identification of novel genes encoding new proteins and the characterization of diseases gene which should provide new insights into mechanisms of disease. Comparative genome mapping will also provide important insights into the evolution of the mammalian genome, its chromosomal architecture and its genes and gene families.

12

SPECIATION

During the 1940s officials in Trinidad launched an intensive campaign to control malaria. They spent much money on spraying and draining marshes, in the belief that malaria was transmitted by *Anopheles albimanus*, a swamp-breeding mosquito that is the principal vector of malaria in Latin America. The campaign failed because the principal vector of malaria in Trinidad was *Anopheles bellator*, a mosquito that breeds in water held within the leaves of bromeliads (relatives of pineapples) growing on the branches of palm trees.

Similarly, in Europe, malaria was believed to be transmitted by mosquitoes of a single species: *Anopheles maculipennis*. European efforts to control malaria sometimes succeeded and sometimes failed, because *A. maculipennis* is not a single species, but consists of at least 18 species that can be distinguished only by examination of their chromosomes. Some of the species breed in fresh water, others in brackish water. Some enter houses; others do not. Furthermore, which mosquito species is the vector of malaria changes regionally. Control efforts are successful only when directed against the species that actually transmits malaria in that area.

Millions of species inhabit Earth. Many of them are easy to identify, but some, such as the species in the *Anopheles maculipennis* complex, are not. All species, living and extinct, are believed to be descendants of a single ancestral species that lived several billion years ago. If speciation had been a rare event, the biological world would be very different than it is today. Speciation is an essential ingredient of evolutionary diversification, and species are the fundamental units of the biological classification systems.

But what are species and how did these millions of species form? How does one species become two? What factors stimulate such splitting? What conditions spur evolutionary radiations? Does speciation accelerate rates of evolutionary change? These and other related questions are the subject of this chapter.

What Are Species?

The word "species" means, literally, "kinds;" but what do we mean by "kinds"? Someone who is knowledgeable about a group of organisms, such as orchids or lizards, usually can distinguish the different species of that group found in a particular area simply by examining them superficially. The patterns of morphological similarities that unite groups of organisms and separate them from other groups are familiar to all of us. The standard field guides to birds, mammals, insects, and flowers are possible only because most species are cohesive units that may change in appearance only gradually over large geographic distances. We can easilyrecognize red-winged blackbirds from New York and red-winged black-birds from California as members of the same species.

But not all members of a species look that much alike. How do we decide whether similar but different individuals should be called different species or regarded as varieties within a species? The concept that has guided these decisions for a long time is genetic integration. If individuals within a population mate with one another but not with individuals of other populations, they constitute a reproductively isolated group within which genes recombine; that is, they are *independent evolutionary units*. These independent evolutionary groups are usually called species. Identifying and naming species correctly is important because when different investigators report the results of studies of a particular species, we need to know that they are writing about the same species.

More than 200 years ago the Swedish biologist Carolus Linnaeus, who originated the system of naming organisms that we use today, described hundreds of species. Because he knew nothing about the mating patterns of the organisms he was naming, Linnaeus classified them on the basis of their appearance. Many species that were classified by their appearance, when nothing was known about their reproductive behavior, are actually independent evolutionary units. This is not surprising because the members of an evolutionary unit share genes inherited from common ancestors. These individuals look alike because they share many of the alleles that code for their structures.

The species definition that has been used by most biologists was proposed by Ernst Mayr in 1940. He stated, "Species are groups of actually or potentially interbreeding natural populations which are reproductively isolated from other such groups." The "groups" in this definition are collections of local populations. The words "actually or potentially" assert that, even if some members of a species are not in the same place and hence are unable to mate, they should not be placed in separate species if they would be likely to mate if they were together. The word "natural" is an important part of the definition because only in nature does the exchange of genes, which occurs within species, affect evolutionary processes; the interbreeding of two different species in captivity does not.

Gene exchange is the main reason why species are cohesive units. Individuals that mate with each other "recognize" one another as suitable mates. Many biologists study how individuals use visual, vocal, and chemical clues to recognize suitable mates and to avoid mating with individuals belonging to other species.

Biologists attempting to reconstruct the evolutionary histories of organisms also try to identify independent evolutionary units.

How Do New Species Arise?

Not all evolutionary changes result in new species. Evolution creates two patterns across time and space: *anagenesis* and *cladogenesis*. *Anagenesis* is change in a single lineage through time. With sufficient time, the changes may be so great that the descendants are given another name, even though no "new" species has formed. Anagenetic changes are a common feature of the fossil record.

Cladogenesis (*speciation*) is the process by which one species splits into two species, which thereafter evolve as distinct lineages. Although Charles Darwin entitled his book *The Origin of Species*, he did not extensively discuss how a single species splits into two or more daughter species. Rather, he was concerned principally with demonstrating anagenesis, that species are altered by natural selection over time.

The critical process in the formation of two species from one ancestral species is the separation of the gene pool belonging to the ancestral species into two separate gene pools. Subsequently, within each isolated gene pool, allele and gene frequencies may change as a result of the action of evolutionary agents. If sufficient differences have accumulated during the period of isolation, the two populations may not exchange genes when they come together again.

Gene flow among populations may be interrupted in several ways, each of which characterizes a mode of speciation. The next three sections focus on these modes of speciation: sympatric speciation, allopatric speciation, and parapatric speciation.

Sympatric Speciation Occurs without Physical Separation

The subdividing of a gene pool even though members of the daughter species are not physically separated during the process is called *sympatric speciation* (*sym-,* "with"; *patris,* "country"). The most common means of sympatric speciation is *polyploidy*, a duplication of the number of chromosomes.

Polyploidy arises in two ways. One way is the accidental production during cell division of cells having four (tetraploid) rather than two (diploid) sets of chromosomes. This process produces an *autopolyploid* individual, one having more than two sets of chromosomes, all derived from a single species. This tetraploid individual cannot mate successfully with diploids, but if it self-fertilizes or mates with other tetraploids, a new evolutionary lineage may form.

A polyploid species can also be produced when individuals of two different species, whose chromosomes do not pair properly during meiosis, interbreed. The resulting individuals, called allopolyploids, are usually sterile, because the chromosomes from one species do not pair properly with those from the other species during meiosis, but they can reproduce asexually. After many generations, some of the individuals may become fertile as a result of further chromosome duplication.

Polyploidy can create a new species among plants much more easily than among animals because plants of many species can reproduce by self-fertilization as well as by crossing with a relatively unrelated individual. If the polyploidy arises in several offspring of a single parent, the siblings can fertilize one another. Speciation by polyploidy has been very important in the evolution of flowering plants. Botanists estimate that more than half of all species of flowering plants are polyploids. Most of these arose as a result of hybridization between two species, followed by selffertilization.

The importance of allopolyploidy and the speed with which it can produce new species are illustrated by salsifies (*Tragopogon*), members of the sunflower family. Salsifies are weedy plants that thrive in disturbed areas around towns. People have inadvertently spread them around the world from their ancestral ranges in Eurasia. Three diploid species of salsify were introduced into North America early in this

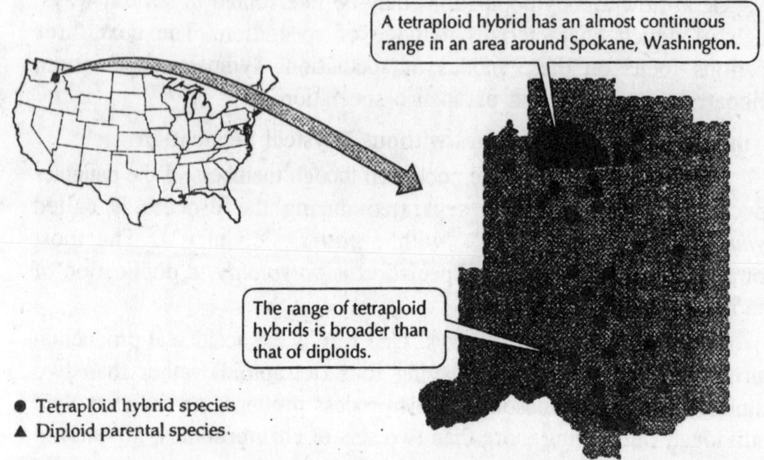

Fig. 12.1. Polyploids can outperform their parents. Tragopogon species are members of the sunflower family. The map shows the distribution of the diploid parent species and the tetraploid hybrid species of Tragopogon in eastern Washington and adjacent Idaho.

century: *T. porrifolius*, *T. pratensis*, and *T. dubius*. Two tetraploid hybrids–*T. mirus* and *T. miscellus* between species of the original three were first reported in 1950. Both hybrids have spread since their discovery and today are more widespread than their diploid parents.

Studies of their cells have revealed that both types of hybrids have been formed more than once. Some populations of *T. miscellus*, for example, have the chloroplast genome of *T. pratensis*, whereas other populations have the chloroplast genome of *T. dubius*. Differences in the ribosomal genes of local populations of *T. miscellus* show that this allopolyploid has evolved independently at least three times. Scientists seldom know the dates and locations of species formation so well. The success of newly formed hybrid species of salsifies illustrates why so many species of flowering plants originated as polyploids.

Among animals, sympatric speciation by polyploidy is relatively rare, but a few cases are known. The tree frog *Hyla versicolor* is a tetraploid species with a broad range in eastern North America. *H. versicolor* arose recently as a result of at least three different hybridizations between individuals from eastern and western populations of its diploid relative *Hyla chrysocelis*. The eastern and western populations of *H. chrysocelis* were isolated from one another for about 4 million years, enough time for substantial genetic differences to accumulate. Today the ranges of the two populations abut in a northsouth line west of the Mississippi River. Eastern and western individuals rarely hybridize in nature, probably because the calls of the males are so different that females respond only to the calls of males of their own type when they have a choice.

Among animals, sympatric speciation as a result of habitat selection is more common than speciation by polyploidy. A good example is speciation in the picture-winged fruit fly, *Rhagoletis pomenella*, in New York. Until the mid 1800s, these fruit flies courted, mated, and deposited their eggs only on hawthorn berries. The larvae learned the odor of hawthorn as they fed on the berries, and when they emerged from theirpupae they used this food-based memory to locate other hawthorn plants. About 150 years ago, large commercial apple orchards were planted in the Hudson River Valley. A few female *Rhagoletis* apparently laid their eggs on apples, perhaps by mistake. Their larvae did not grow as well as the larvae on hawthorn berries, but many did survive. These larvae had learned the odor of apples, so when they emerged as adults they sought out apple trees, where they mated with other flies reared on apples. Today there are two sympatric species of *Rhagoletis* in the Hudson River Valley. One feeds on hawthorn berries, the other on apples. The two species are reproductively isolated because they mate only with individuals raised on the same fruit, and because they emerge from their pupae at different times. In addition, apple-feeding flies have evolved so that they now grow more rapidly on apples than they originally did.

Allopatric Speciation Requires Isolation by Distance

Speciation that results when a population is divided by a barrier is known as *allopatric speciation* (*allo-*, "different"), or *geographic speciation*. Allopatric speciation is the dominant form of speciation among most groups of organisms. The range of a species may be divided by a physical barrier, such as water gaps for terrestrial organisms, dry land for aquatic organisms, and mountains. Barriers can form when continents drift, sea levels rise, or climates change. Populations separated in this way are often large initially. They evolve differences because the places in which they live are or become different.

Alternatively, allopatric speciation may result when some members of a population cross an existing barrier and found a new isolated population. Populations established in this way are typically small at first. They usually differ genetically from their parent populations because a small group of individuals has only an incomplete representation of the genes found in its parent population.

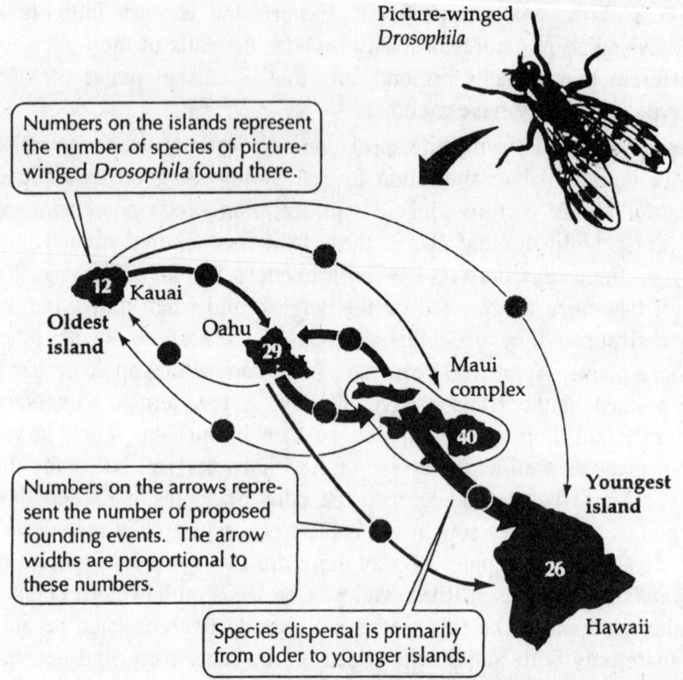

Fig. 12.2. Founder events lead to allopatric speciation. The extremely high level of speciation found among Drosophila in the Hawaiian island is almost certainly the result of founder events-new populations founded by individuals dispersing among the islands.

Allopatric speciation by this sampling effect is illustrated by the singing hone-yeater, a common Australian bird that lives on the mainland and on coastal islands. The birds on Rottnest Island, 20 km off the coast of Western Australia, sing fewer song types than main land birds do, and their songs have fewer syllables and notes. Evidently the island colonizers did not carry all of the alleles responsible for the full range and complexity of songs found in mainland individuals. Mainland singing honeyeaters do not respond to the songs of island birds and Rottnest birds do not respond to songs of mainland birds.

Dispersal across barriers often leads to species formation. For example, many of the hundreds of species of the fruit fly *Drosophila* in the Hawaiian Islands are restricted to a single island. They are almost certainly the result of new populations founded by individuals dispersing among the islands, because the closest relative of a species on one island is often a species on a neighboring island rather than a species on the same island. On the basis of studies of their chromosomes, speciation among the picture-winged *Drosophila* is believed to have been caused by at least 45 founder events.

If the environments on the two sides of the physical barrier differ, evolutionary agents may cause the populations to diverge further genetically. Differences that accumulate while a barrier is in place may become so large that the populations will fail to establish gene flow if the barrier later breaks down; that is, they will have evolved to be different species.

The finches of the Galapagos Islands, 1,000 km off the coast of Ecuador, demonstrate the importance of geographic isolation for speciation. Darwin's finches (as they are usually called, because Darwin was the first scientist to study them) arose on the Galapagos Islands by speciation from a single South American species that colonized the islands. Today there are 14 species of Galapagos finches, all of which differ strikingly from the probable mainland ancestor.

The islands of the Galapagos Archipelago are sufficiently isolated from one another that the finches seldom migrate between them. Also, environmental conditions differ among the islands. Some are relatively flat and arid; others have forested mountain slopes. Populations of finches on different islands have differentiated enough that when occasional immigrants arrive from other islands, they either do not breed with the residents, or if they do, the resulting offspring usually do not survive as well as those produced by pairs composed of island residents. The genetic distinctness of different populations is thus maintained.

A barrier's effectiveness at preventing gene flow depends on the size and mobility of species. What is a firm barrier for a terrestrial snail may be no barrier at all to a butterfly or a bird. Populations of wind-pollinated plants are totally isolated at the maximum distance pollen is blown by the wind, but individual plants are effectively isolated at much shorter distances. Among animal-pollinated plants, the width of the barrier is the distance that pollinators travel while carrying pollen. Even animals with great powers of dispersal are often reluctant to cross narrow strips of unsuitable habitat. For animals that cannot swim or fly, narrow water-filled gaps may be effective barriers.

Parapatric Speciation Separates Adjacent Populations

Sometimes reproductive isolation develops among adjacent members of a population in the absence of a geographic barrier. Known as *parapatric speciation* (*para,* "beside"), this type of speciation is much less common than allopatric or sympatric speciation because gene flow usually prevents differentiation between populations in contact. Occasionally, however, a species boundary forms where there is a marked change in environment. That is, allopatric speciation becomes parapatric speciation when the geographical boundary separating species becomes extremely small.

Unusually abrupt changes in soil are created by mining activities that leave rubble (tailings) with high concentrations of heavy metals that are detrimental to plant growth. For example, the soils developing on the tailings at the Goginian lead mine near Aberystwyth, Wales, are highly contaminated with lead, but where the tailings end, they suddenly give way to normal rich pastureland.

The pasture grass *Agrostis tenuis* is common on both types of soils, but there is a sharp gradient in lead tolerance among plants less than 20 m apart. Plants on the mine tailings grow well in lead concentrations that would be lethal to plants growing just a few meters away. Nearly complete reproductive isolation exists between plants on contaminated and normal soil because they flower at different times. These two populations have not yet been designated as separate species, but reproductive isolation between them has already evolved, demonstrating that gene flow can slow or stop even in the absence of a distinct physical barrier.

Reproductive Isolating Mechanisms

Once a barrier to gene flow is established, by whatever means, the daughter populations may diverge genetically because of the action of evolutionary agents. Over many generations, differences that reduce the survivability of hybrid offspring may accumulate. In this way, reproductive isolation can evolve as an incidental by-product of other genetic changes in allopatric populations. For example, individuals in the two daughter populations may become so different that they are not recognized as suitable mates, as shown by the singing honeyeaters on Rottnest Island.

However, geographic isolation does not necessarily mean reproductive isolation. For example, American and European sycamores have been physically isolated from one another for at least 20 million years.

Nevertheless, they are morphologically very similar, and they can form fertile hybrids. They lack traits that would prevent individuals of two different populations from producing fertile hybrids. In this section we examine the ways in which such traits—reproductive isolating mechanisms—arise. Then we explore what happens when reproductive isolation is incomplete.

Prezygotic Barriers Operate before Mating

Individuals of different species may select different places in the environment in which to live. As a result, they may never come into contact during their respective mating seasons; that is, they are reproductively isolated by location (*spatial isolation*). Many organisms have mating periods that are as short as a few hours or days. If the mating periods of two species do not coincide, they will be reproductively isolated by time (*temporal isolation*). Differences in the sizes and shapes of reproductive organs may prevent the union of gametes from different species (*mechanical isolation*). Sperm of one species may not be attracted to the eggs of another species because the eggs do not release the appropriate attractive chemicals, or the sperm may be unable to penetrate the egg because it is chemically incompatible (*gametic isolation*). These mechanisms, all of which operate before mating, are called *prezygotic reproductive barriers*.

Postzygotic Barriers Operate after Mating

If individuals of two different populations still recognize one another and mate, *postzygotic reproductive* barriers may prevent gene exchange. Genetic differences that accumulated while the daughter populations were allopatric are likely to reduce the survivability of offspring produced by matings between individuals from the two daughter populations.

The offspring of parents from genetically dissimilar populations are known as hybrids. Hybrid zygotes may be abnormal (*hybrid zygote abnormality*), or the hybrids may mature normally but be infertile when they attempt to reproduce (*hybrid infertility*). For example, the offspring of matings between horses and donkeys-mules—are vigorous, but mules are sterile; they produce no descendants.

If hybrid offspring are less viable than offspring resulting from matings within populations of the daughter species, postzygotic barriers may be reinforced by the evolution of more effective prezygotic barriers. More effective prezygotic barriers should evolve because individuals engaging in hybrid matings leave fewer surviving offspring than those that mate only within their group. Reinforcement of prezygotic barriers has been demonstrated in a few laboratory populations, but evidence for it in nature has been slow to accumulate.

Sometimes Reproductive Isolation is Incomplete

Sometimes contact is reestablished between formerly geographically isolated populations before many genetic differences have accumulated. Then the individuals may interbreed freely with members of the other population and produce hybrid offspring that are as successful as those resulting from matings within each population. If successful hybrids spread through both populations and reproduce with other individuals, the gene pools combine quickly, and no new species result from the period of isolation.

Alternatively, rather than thoroughly combining their gene pools, the two populations may interbreed only where they come into contact, resulting in a hybrid zone. To determine what happens when formerly separate populations come together, studies ideally begin when contact is first established. Blue and snow geese provide an opportunity to observe the formation of a hybrid zone. These geese breed in Arctic North America and spend the winter in the southern United States. Birds with white plumage (snow geese) dominate breeding populations in the West; birds with dark plumage (blue geese) dominate in the East. Historical evidence shows that the two color forms were almost completely separated geographically until the 1930s.

The recent hybrid zone has resulted from a change in the winter feeding ranges of the birds. Birds of both types now winter in large flocks in the rice-growing regions of inland Texas and Louisiana. The geese select mates while on the wintering grounds, and pairs migrate to nest on the breeding grounds from which the female came. Interbreeding is now common between the two forms, and a hybrid zone is developing in a small region of the Canadian Arctic. Biologists are monitoring the spread of this hybrid zone to determine whether isolating mechanisms are developing or whether the zone will continue to spread.

We can measure genetic differences among species

If two species hybridize, we know that they are very similar genetically, but the absence of interbreeding tells us nothing about how dissimilar two species are. Not until modern molecular tools were developed could biologists measure genetic differences among species. Molecular studies are now demonstrating that many sympatric species differ from one another very little genetically. For example, flies of different species of Hawaiian *Drosophila* share nearly all of their alleles. Most morphological differences among the species are based on variability already present within each of the species. All of the hundreds of species of this genus that have evolved in Hawaii during the past 40 million years, even those that have diverged morphologically, are relatively similar genetically. Other research confirms that the differences among species generally are similar in type to the differences within species.

Variation in Speciation Rates

Some lineages of organisms have many species; others have few. The hundreds of species of *Drosophila* found in the Hawaiian Islands have evolved within the last 40 million years. In contrast, there is only one species of horse-shoe crab, even though its lineage has survived more than 200 million years. Why do rates of speciation vary so widely among lineages? In the sections that follow we will examine several factors that influence speciation rates: species diversity and range size, life history traits, environment, and generation times.

Species Richness may Favor Speciation

The larger the number of species, the larger the number of opportunities for new species to form. This is particularly true of speciation by polyploidy because more species are available to hybridize with one another. It is also partly true for geographic speciation, because the number of ranges bisected by a given barrier should be positively correlated with the number of species living in the area.

However, the relationship between range size and speciation rate is not simple, because ranges of individual species tend to be smaller where there are many species. The larger the range of a species, the more likely a physical barrier is to subdivide it. Conversely, the smaller the range size, the less likely a particular randomly placed barrier will subdivide it. Also, species with large ranges are more likely than species with small ranges to establish isolated peripheral populations that survive long enough to form new species.

Random variation in rates of events that create barriers is an important cause of variable speciation rates. Where and when geographic barriers arise, and where and when genetic accidents that result in polyploid individuals happen, are unpredictable and variable. Nonetheless, traits of species may influence how often their ranges are divided by barriers.

Life History Traits Influence Speciation Rates

Individuals of species with poor dispersal abilities are unlikely to establish new populations by dispersing across barriers, and even narrow barriers are effective among species whose individuals are highly sedentary. Populations of land snails, which have speciated profusely on Pacific islands, are separated by barriers as narrow as a city street.

Animals with complex behavior are likely to speciate at a high rate because they make sophisticated discriminations among potential mating partners. They distinguish members of their own species from members of other species, and they make subtle discriminations among members of their own species on the basis of size, shape, appearance, and behavior. These discriminations may be based on the quality of the genes of the potential partner, the quality of parental care likely to be given, or both. Such behavioral discrimination can greatly influence which individuals are most successful in producing offspring.

Therefore, mate selection is probably a major cause of rapid evolution of reproductive isolation between species.

Heterogeneous Environments Favor High Speciation Rates

Speciation rates among different lineages of the large, hoofed mammals of Africa are correlated with diet: Grazers and browsers speciate faster than omnivores and anteaters.

The grazers and browsers require large expanses of open grassland and woodland, respectively. In Africa, these resources disappeared from and reappeared in large areas during periods of climatic change, thus isolating populations and causing both high extinction rates and high rates of differentiation among populations between these isolated regions. Omnivores and anteaters, on the other hand, maintained more continuous populations during these climatic changes. Gene flow continued among their populations, and reproductive isolation was not established.

Short Generation Times Enhance Speciation

We have concentrated on factors that influence rates at which the ranges of species are subdivided by barriers. But the rate at which new species form also depends on how fast daughter populations diverge. The more rapidly they diverge, the sooner they are likely to evolve reproductive isolating mechanisms and the less likely they are to hybridize when they again become sympatric. Shorter generation times result in more generations per unit time and, as a result, generate the potential for more evolutionary changes per unit time.

Evolutionary Radiations: High Speciation and Low Extinction

The fossil record reveals that, at certain times in some lineages, speciation rates have been much higher than extinction rates. The result is an *evolutionary radiation* giving rise to a large number of daughter species. What conditions cause speciation rates to be much higher than extinction rates?

Evolutionary radiations are likely when a population colonizes an environment that has relatively few species. This condition typifies islands because many organisms disperse poorly across large water-filled gaps. Because islands lack many plant and animal groups found on the mainland, ecological opportunities exist that may stimulate rapid evolutionary changes. Water barriers also restrict gene flow among islands in an archipelago, so populations on different islands can evolve adaptations to their local environments. Together these two factors make it likely that speciation rates will exceed extinction rates.

Remarkable evolutionary radiations have occurred in the Hawaiian Archipelago, the most isolated islands in the world. The Hawaiian Islands lie 4,000 km from the nearest major land mass and 1,600 km from the nearest group of islands. The islands are arranged in a line of decreasing age—the youngest islands to the southeast, the oldest to the northwest.

The native biota of the Hawaiian Islands includes 1,000 species of flowering plants, 10,000 species of insects, 1,000 land snails, and more than 100 birds. However, there were no amphibians, no terrestrial reptiles, and only one native mammal—a bat—until humans introduced additional species. The 10,000 known native species of insects on Hawaii are believed to have evolved from only about 400 immigrant species; only seven immigrant species are believed to account for all the native Hawaiian land birds.

More than 90 percent of all plant species on the Hawaiian Islands are endemic—that is, they are found nowhere else. Several groups of flowering plants have more diverse forms and life histories on the islands and live in a wider variety of habitats than do their close relatives on the mainland. An outstanding example is the group of sunflowers called silverswords and tarweeds (the genera *Argyroxiphium*, *Dubautia,* and *Wilkesia*). Chloroplast DNA data show that these species share a relatively recent common ancestor, which is believed to be a species of tarweed from the Pacific coast of North America.

Whereas all mainland tarweeds are small, upright, nonwoody plants (herbs), Hawaiian silversword species include prostrate and upright herbs, shrubs, trees, and vines. They occupy nearly all the habitats of the islands, from sea level to above timberline in the mountains. Despite their extraordinary diversification, however, the silverswords have differentiated very little genetically. In other words, the rate of morphological evolution has been much more rapid than the rate of evolution of chloroplast DNA in these plants.

The island silverswords are more diverse in size and shape than the mainland tarweeds because the colonizers arrived on islands that had very few plant species. In particular, there were few trees and shrubs because such large-seeded plants only rarely disperse to oceanic islands; many island trees and shrubs have evolved from nonwoody ancestors. On the mainland, however, tarweeds have lived in ecological communities that contain tree and shrub lineages older than their own—that is, where opportunities to exploit the tree way of life were already preempted.

Evolutionary lineages may also radiate when they acquire a new adaptation that enables them to use the environment in new and varied ways. For example, ancestors of the 94 species of American blackbirds evolved powerful muscles for opening their bills. These muscles enable the birds to obtain food by opening their bills forcibly against objects they wish to move, exposing otherwise hidden prey. Such activity is called *gaping*. Birds lacking these powerful muscles can find prey only on exposed surfaces of objects. Blackbirds gape into wood, fruits, leaf clusters, and stems of nonwoody plants; under sticks, stones, and animal droppings; and into the soil. With this feeding method, they have come to occupy nearly all habitat types in North and South America, and they are among the most abundant birds throughout the region.

Speciation and Evolutionary Change

Does speciation stimulate evolutionary change? In 1972 Niles Eldredge and Stephen Jay Gould proposed that most evolutionary changes take place at the time of speciation. They suggested that the isolation of small, peripheral populations was the most common event leading to rapid evolutionary changes. Their reasoning was that small founding populations lack some of the alleles found in the source populations and have different allele frequencies as a result of random genetic changes. These founding populations might change rapidly because they live in an environment that differs from the one from which they came. According to Eldredge and Gould, the speciation process stimulates evolutionary change, and once speciation has been completed, the better-integrated new genotypes resist change, leading to stasis—long periods of little change that are interrupted only by the next round of speciation. This pattern of evolution is called *punctuated equilibrium*.

The fossil record reveals examples of both stasis and long-term gradual evolution, but it does not tell us if periods of rapid change usually accompany times of speciation and are rare at other times. The fossil record is of limited help in testing this hypothesis because the ages of most fossils cannot be determined precisely enough. Molecular tools, on the other hand, enable evolutionary biologists to measure correlations between speciation and evolutionary change.

Lungless salamanders have been studied extensively at both molecular and morphological levels. The information from these studies allows us to compare the amount of genetic difference with the amount of morphological differences among these species to determine if speciation typically has been accompanied by morphological change. The results show that most speciation events are accompanied by almost no morphological changes. In these salamanders, speciation has proceeded at a much higher rate than morphological evolution. How often the rate of speciation exceeds the rate of morphological evolution is unknown.

The powerful gaping muscles of the meadowlark, one of the American blackbirds, enable it to expose prey buried in soil.

Significance of Speciation

The result of speciation processes operating over billions of years is a world in which life is organized into millions of species, each adapted to live in a particular place and to use environmental resources in a particular way. Earth would be very different if speciation had been a rare event in the history of life. How the millions of species are distributed over the surface of Earth and organized into ecological communities will be a major focus of Part Seven of this book, "Ecology and Biogeography," at which time we will also discuss how human activities are causing the extinction of many species and what we can do to reduce the rate of species loss.

INDEX